Weather Cycles
Second edition

This completely updated new edition of *Weather Cycles: Real or Imaginary?* explores in detail the unresolved debate on the existence of weather cycles. The book examines the competing arguments for observed effects being due to natural variability, solar activity and the Earth's orbital parameters.

A wide range of events are presented, from ice ages to various oscillations, including El Niño, and other examples of apparently cyclic behaviour, drawing on instrumental observations and other records such as tree rings, ice cores, ocean sediments, corals and stalagmites. The book provides the basic statistical analysis and climatic theories of non-linear systems (chaos theory) to assess the data. The conclusion is that, with few exceptions, the case for weather cycles is not proven, but that an appreciation of the apparently periodic nature of climatic fluctuations is essential in understanding contemporary changes in the Earth's climate.

Weather Cycles: Real or Imaginary? provides a different perspective on one of the most difficult questions in the current global warming debate: namely, just how much of the recent temperature rise can be attributed to natural causes? Only by understanding how the climate can change of its own accord, and whether observed shifts are part of a set of predictable patterns, will it be possible to reach a reliable judgement on how much impact human activities are having. This book examines the complex analysis required to assess the evidence for cycles with a minimum of mathematics. This comprehensive and balanced account will appeal to the student and expert alike.

After 7 years at the UK National Physical Laboratory researching atmospheric physics, **Bill Burroughs** spent 3 years as a UK Scientific Attaché in Washington DC. Between 1974 and 1995, he held a series of senior posts in the UK Departments of Energy and then Health. He is now a professional science writer and has published several books on various aspects of weather and climate (two as a co-author), and three books for children on lasers. These books include *Watching the World's Weather* (1991), *Weather Cycles* (1992), *Does the Weather Really Matter?* (1997), *The Climate Revealed* (1999), and *Climate Change: A Multidisciplinary Approach* (2001), all with Cambridge University Press. In addition, he acted as lead author for the World Meteorological Organization on a book entitled *Climate: Into the Twenty-First Century*. He has also written widely on the weather and climate in newspapers and popular magazines.

Weather Cycles

Real or Imaginary?
(Second edition)

William James Burroughs

CAMBRIDGE
UNIVERSITY PRESS

PUBLISHED BY THE PRESS SYNDICATE OF THE UNIVERSITY OF CAMBRIDGE
The Pitt Building, Trumpington Street, Cambridge, United Kingdom

CAMBRIDGE UNIVERSITY PRESS
The Edinburgh Building, Cambridge, CB2 2RU, UK
40 West 20th Street, New York, NY 10011–4211, USA
477 Williamstown Road, Port Melbourne, VIC 3207, Australia
Ruiz de Alarcón 13, 28014 Madrid, Spain
Dock House, The Waterfront, Cape Town 8001, South Africa

http://www.cambridge.org

First edition 1992

Second edition 2003

Printed in the United Kingdom at the University Press, Cambridge

Typefaces Utopia 9.25/13.5 pt. and Meta Plus   *System* LaTeX 2ε   [TB]

A catalogue record for this book is available from the British Library

ISBN 0 521 82084 7 hardback
ISBN 0 521 52822 4 paperback

Contents

Preface

When I wrote the preface to the first edition of this book over 10 years ago, I noted that the history of meteorology was littered with whitened bones of claims to have demonstrated the existence of reliable cycles in the weather. These failures had led many to conclude that the search for cycles is a pointless exercise. Not much has changed over the intervening years, but the pace of effort to provide better answers about the periodic behaviour of the climate has quickened.

The reasons for what might be regarded a Sisyphean task are not hard to find. Now more than ever, we need to understand why the climate changes and to what extent human activities are producing effects which are comparable to or exceed the natural variability of the global climate. The record-breaking warmth of global temperature in the 1990s has made it even more obvious that only by understanding the true nature of climatic fluctuations will it be possible to reach an early conclusion on what proportion of current global warming is due to natural causes. This sense of urgency was built on the emerging realisation during the 1980s that quasi-cyclic fluctuations in the tropical Pacific (El Niño) had global implications. When the record-breaking El Niño of 1982/83 was followed by a comparable event in 1997/98 there was widespread speculation that these events were in some way linked to global warming and hence to anthropogenic activities.

This book sifts through the huge amount of work that has been published on weather cycles. The aim is to clear the air by identifying the consistent features in the climate records and providing a basis for addressing the continuing stream of new evidence of periodicities in the weather. This is no easy matter, for although the work to squeeze evidence for cycles out of weather records seems to have produced so little, recent rapid progress in exploring the possible causes of quasi-periodic changes in the global climate has stimulated widespread public and scientific interest. Building on the scientific studies during the 1980s of the global impact of El Niño, a whole new 'oscillations' business has grown up to explore how the atmosphere and the oceans combine to produce approximately regular fluctuations in regional climates.

At the same time the awe-inspiring complexity of the climate has been underlined by the emergence of chaos theory. This new approach to non-linear systems has changed our view of the climate. These altered perceptions have been reinforced in recent years by both ice core and ocean sediment records showing that during the last ice age conditions fluctuated much more wildly that in recent millennia. When combined with new theories that suggest ocean currents can flip between radically different states, there is now a much greater awareness of the unpredictable nature of the climate. On top of all this, computer models suggest that the climate can exhibit approximately regular fluctuations that are essentially chaotic. So, is it any wonder that these developments appear to endorse the frustration of researchers who have toiled so long in a largely vain effort to establish the existence of weather cycles?

The real value of all this work is, however, that it provides insight into how the Earth's climate, driven by the annual orbit round the Sun, hovers on the edge between order and disorder. This means that, while we should not have too high expectations of what the search for weather cycles may produce, there are good reasons for looking for semblances of order. Only by grappling with the wealth of often contradictory information can we develop a clearer picture of the causes of climatic change. And, as the world appears destined to enter a period of warming beyond anything in recorded history, this picture is of vital importance to us all. As the pace of the debate quickens, this book seeks to provide the basis for addressing these unanswered questions so that we can avoid jumping to the rash conclusions that have bedevilled the search for weather cycles in the past.

Acknowledgements

As I noted in the first edition, I owe a particular debt of gratitude to the late J. Murray Mitchell, Jr, who over many years offered me invaluable advice and guidance on weather cycles, drawing on his unparalleled knowledge and insight of this subject. This early inspiration has continued to help me in the task of reviewing the gathering pace of work in this field. I have also been helped considerably by discussions with Keith Briffa, Chris Folland, Giles Harrison, Phil Jones, Theordor Landscheit, Judith Lean, Jan Lindstrom, Leslie Malone, David Parker and Tony Slingo, information provided by Sallie Baliunas, John Donnarummo, Per Kallberg, Markus Kunze, Karin Labitzke, Robert Livezey, Michael Ram, Eugene Rasmussen and Brian Tinsley, and editorial advice from Martin Hoyle. Finally, I have to thank my wife who has supported me unstintingly over many years with this book and associated studies.

Chapter 1

The search for cycles

And the seven years of plenteousness, that was in the land of
Egypt was ended. And the seven years of dearth began to
come, according as Joseph had said: and the dearth was in all
lands; but in the land of Egypt there was bread.

Genesis 41:53

Throughout recorded history the fluctuations of the weather have played a
major part in human life. Times of feast and times of famine have repeatedly
occurred. The biblical story of Joseph's dream, accurately foretelling that
7 good years would be followed by 7 years of famine and describing the
action that was taken to store the surplus from the good years to meet the
shortages of the bad years appears to be the first recorded example of a
periodic variation in the weather over a number of years, but it also shows
the huge benefit that can accrue from the accurate predictions of such
regular meteorological changes and their impact on harvests, and explains
why the possibility of regular fluctuations in the weather has fascinated
weather watchers for so long.

There may also be a more fundamental reason for searching for such
orderly behaviour in the weather. Because so much of our lives is governed
by the rhythms of the seasons, it is natural to look for the same sort of order
in the longer term, more chaotic behaviour of the physical world around us.
Nowhere is this desire for order more widely expressed than in those who
attempt to explain fluctuations from year to year in the weather. The daily
and annual progression of the weather is dominated by the predictable
rotation of the Earth and its motion round the Sun, and it is therefore
natural to ask whether the other fluctuations, which are such a feature
of our weather, have a simple explanation.

We all know that the weather is rarely, if ever, behaving normally.
Climatology textbooks can tell us what, on the basis of long-term records,

the average conditions are for any given place at any given time. But, in practice, it is almost always hotter or colder, or wetter or drier than these normals. Over periods of weeks or months these fluctuations may add up to give a notable cold spell, heatwave or drought. The occurrence of such extremes is a source of constant fascination for meteorologists. They appear on every timescale, from week to week, month to month, year to year and over the decades and centuries. Over all these periods the weather appears to behave in a chaotic way that defies description. Yet we all intuitively suspect that there is some underlying order. Extreme spells of weather seem to be balanced out by the opposite extreme with monotonous regularity. As Wiltshire folklore states:

> There is no debt so surely met
> As wet to dry and dry to wet.

On the longer timescale there is widespread assumption that, say, cold winters or hot summers come every so many years. The general public tends to accept that such patterns exist and that the application of suitable scientific analysis will find the key to unlock the door to long-term predictability of the weather.

Among the meteorological community the debate continues as to whether patterns exist and, if so, whether they either constitute a sufficiently large proportion of the observed variability or are sufficiently well established to provide the physical basis for forecasting. This uncertainty exists, in spite of a huge amount of work over many years. The history of this search, how patterns have been detected and what they tell us about the balance between order and chaos in the weather are the themes of this book.

1.1 Social and economic preamble

The story from the Book of Genesis shows that the social and economic implications of major weather fluctuations are profound. Since the advent of reliable instrumental records it has been possible to make estimates of the extent to which various aspects of economic activity have been influenced by abnormal weather events. This provides a basis for making some observations about the potential benefits that might accrue from being able to anticipate periodic fluctuations in the weather. Conversely, apparently regular variations in past economic indicators, such as European cereal prices, may make it possible to draw some inferences about past climatic fluctuations. In theory, this is a practical proposition as such records exist for several hundred years before instrumental records began. Moreover, they can be compared with other indirect records such as measurements of the width of tree rings and wine harvest dates, which have also been obtained in the same area over the same period (see also Chapter 4). So it helps to set the scene by considering what the social and economic implications of weather cycles might be.

The importance of identifying predictable cyclic behaviour in the weather can be gauged from recent events that show some evidence of periodic behaviour. The most celebrated of these is El Niño. This phenomenon, which involves major shifts in both the atmospheric pressure patterns and sea surface temperatures (SSTs) over a large part of the tropical Pacific, occurs every few years. In both 1982/83 and 1997/98 it had a major global impact. In particular, the first of these two events was associated with major droughts in Australia, many parts of sub-Saharan Africa, Brazil and Central America. These extremes inspired a great deal of research, which has provided an increasingly clear measure of how variations in the sea surface temperature play a part in climatic fluctuations around the world. So, if an adequate physical explanation were produced to explain these approximately regular fluctuations, the potential forecasting value of such understanding could have to global economic and social implications.

Similar arguments apply to cold winters in industrial countries. In January 1977, the eastern United States almost ground to a halt. The intense cold precipitated an energy emergency and the total economic cost of the disruption was estimated in 1977 prices to be nearly $40 billion. Subsequent studies suggested that there is an as yet unexplained link between the 11-year cycle in the variability of the Sun and winter temperatures in the south-eastern United States. The link was, however, a complicated one that involved both solar variability and a periodic reversal of the winds in the stratosphere over the equator. Although the value of this proposed connection has not stood the test of time, it is an interesting example of the type of apparently cyclic behaviour that continues to hint at some underlying order in weather patterns. But to have any value, such putative periodic behaviour must be put onto a reliable scientific footing so it can become the foundation of weather forecasts months in advance. Then the potential economic importance of being able to plan, say, energy supplies to accommodate extreme winters will be huge.

Even more important in terms of economic consequences are the cycles of drought that seem to afflict the great plains of the United States. Ever since the dust-bowl years of the 1930s, there has been intense speculation about the existence of an approximately 20-year cycle in rainfall. Subsequent dry periods in the 1950s and again around 1980 reinforced these claims, although it is not yet clear whether the predicted drought around 2000 has truly materialised. The areal extent and the timing of these droughts do not follow a simple pattern, but the implications for US agriculture are clearly substantial. Moreover, because surplus cereal production in the United States has traditionally played a dominant role in meeting shortfalls elsewhere around the world, this behaviour has global consequences.

Similar observations can be made about the economic impact of weather events in the UK and across Europe. The severe winters of 1947, 1963 and 1979 all caused major economic disruption. By the same token, the summer of 1976 demonstrated that even the UK can suffer damaging droughts. In England there is a tendency for hot dry summers to occur every 13 years or so and this provides another hint of underlying periodicity.

Although there is no doubt about the economic impact of weather fluctuations, the converse exercise of seeking to extract information about weather cycles from some economic series is fraught with difficulties. A foretaste of these pitfalls can be seen in Fig. 1.1. This shows that between 1529 and 1541, the thickness of tree rings in oaks in Germany showed a remarkably consistent alternation between thick and thin rings in successive years. This suggests a run of alternating good and bad growing seasons: an inference that is supported by data for the dates of wine harvests (a measure of the quality of the harvest), which are remarkably in step. In contrast, the price of cereals, as measured in a variety of market towns across Europe, does not show any close parallelism, in spite of the fact that the weather probably produced significantly different harvests in each year.

This *hors d'oeuvre* shows the fascinating information that can be extracted from a variety of historic sources both to examine evidence of climatic change and to search for weather cycles. But, do not assume everything is going to be plain sailing. There are a number of snags in the beguiling curves in Fig. 1.1. First, the link between tree-ring width and the weather is complicated. Although hot dry growing seasons tend to produce thin rings and cool wet years produce thick rings, the relationship is by no means simple. Tree-ring width does show changes throughout the growing season but may also be influenced by groundwater reserves from earlier wet seasons. In fact, the wine harvest dates provide a better measure of the weather during the period April to September, as well as being a useful guide to the economic impact of the weather over the period. Second, the behaviour of the cereal prices shows the problems of moving further away from direct meteorological measurements. What must be remembered is that meteorology will be only part of the story. Demographic pressures, civil unrest and other social changes all played a significant part in cereal

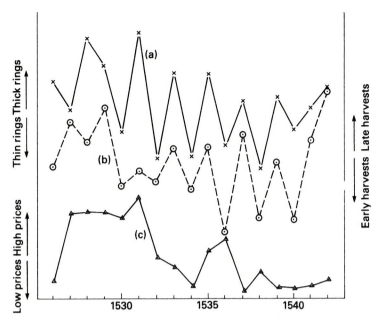

Fig. 1.1. Examples of: (a) German tree-ring thicknesses, (b) French wine harvest dates, and (c) European wheat prices year by year between 1526 and 1542. Tree-growth and wine harvest dates showed a marked biennial oscillation during the 1530s.

prices at the time. Third, and perhaps most important, the splendidly regular fluctuation of tree-ring width is the best example of such a 'biennial oscillation'. Elsewhere, the record is much less regular. This identifies a fundamental weakness of many apparently convincing examples of 'weather cycles': they come and go in a most tantalising manner.

These words are a warning for what will follow. Wherever efforts are made to identify the existence of weather cycles, the form of the original data must be subject to critical scrutiny. This is central to examining meteorological data, and is even more necessary when attributing cycles in economic series to underlying fluctuations in the weather. Moreover, it is of paramount important when going so far as to estimate the economic consequences of predicted periodic variations in the weather. Failure to exercise this critical faculty can lead to economic nonsense. This is a discipline that many cycle enthusiasts have not always maintained in their efforts to promote the case for their discoveries.

So much for economics. We must now turn to the case for the cyclic behaviour of the weather, starting with a brief history of the early attempts to explain apparently periodic variations in the weather. This will set the scene for describing the mathematics and science of making reliable investigations of meteorological data and also the latest work on developing the case for and against weather cycles. Behind all this work lies the knowledge that if it could be established why the weather should fluctuate in a regular and predictable manner, the economic benefits would be potentially vast.

1.2   History of cycle-searching

Apart from the Book of Genesis, Theophrastus of Eresus, Lesbos, made the first recorded observation of weather cycles in the fourth century BC. He was a younger friend of Aristotle, studied at Plato's Academy, and became Aristotle's chief assistant after Plato's death. Together, they made a study of the whole of nature, with Aristotle taking animals and Theophrastus taking plants. In his study of meteorology he noted that 'the ends and the beginnings of the lunar month are apt to be stormy'. Over 2000 years later the debate still continues about the extent of lunar effects on the weather.

The ancient art of defining patterns in weather, which is encapsulated in folklore, was mentioned earlier. Frequently, these patterns are concerned with month-to-month, or season-to-season variations. Only rarely do these rules extend to changes from year to year. Because the central concern in this book is periodicities longer than a year, it is these more speculative saws that are of more interest. In this context the following example is intriguing:

> Extreme seasons are said to occur from the sixth to tenth year of each decade, especially in alternating decades.

This suggests the detection of periodicities of around 10 and 20 years. As will be seen, these figures are close to two of the most thoroughly studied weather cycles.

There is little evidence that prior to the Age of Reason there was any attempt to quantify the variations. One interesting exception appears to be the 35-year rhythm, which, according to Francis Bacon, was already a subject of inquiry in the Low Countries at the beginning of the seventeenth century. This periodicity was to gain much greater attention in the late nineteenth century when the Swiss Professor E. Bruckner was commissioned by the Russian Government to study changing levels in the Caspian Sea, which caused dislocation of transport. He investigated weather data from all over Europe for rivers, lakes, harvests and vintages and concluded that there was a 35-year cycle affecting weather, and thus the changing levels in the Caspian Sea. As we will see, after languishing in obscurity for much of the twentieth century, this periodicity has re-emerged as a feature in tree-ring studies that may be linked to quasiperiodic fluctuations of the atmosphere–ocean interactions in the North Atlantic. In more scientific studies, the first example of seeking to explain weather variations was by the astronomer William Herschel in the early nineteenth century. He proposed that the changes in the Sun's output could influence the weather. But it was the work of another astronomer that truly set in motion the subject of solar cycles in the weather. In 1844, Heinrich Schwabe discovered that the number of sunspots varied in a regular, predictable way,[1] leading to scientific speculation that our weather could vary in the same pattern.

A measure of the increasing rate of the search of weather records for cycles and hidden periodicities is in Sir Napier Shaw's classic manual of meteorology, published between 1926 and 1932,[2] which noted more than 100 cycles that had been 'discovered'. The complexity of these investigations, their possible implications and underlying weaknesses are neatly encapsulated by a quotation in his more popular book on the drama of the weather:

> The lunar–solar cycle of 744 years has been invoked by Abbé Gabriel. It combines 9202 synodic revolutions, 9946 tropical, 9986 draconitic, 9862 anomalistic, 40 revolutions of the ascending node of the lunar orbit and 67 periods of sunspots. It has harmonics of 372 years, 186 years. The last was relied upon for a prediction, made in the summer of 1925, of a cold winter to follow. The prediction was fulfilled in England by the occurrence of exceptionally cold weather in November, December and January. It must, however, be remarked that February, which is accounted as a winter month, brought the highest recorded temperature of that month for 154 years, and a spell of weather compared with which the first half of May was wintery.[3]

Another example of periodicity cited by Sir Napier Shaw shows the problems of obtaining a close correlation with sunspots over a limited period. The example he gave was of an apparent link between the level of Lake Victoria and sunspots over the period 1902 to 1921 (see Fig. 1.2). Despite the strength of this association, the prediction of a high level of the lake with the next sunspot maximum in 1928 proved incorrect. Subsequently, the low levels of the lake occurred every 5 years or so, and also the range of variation in lake level reduced. Even more important, it is now known there have been bigger and more lasting changes in the lake level unconnected with solar activity. First, a decline of nearly 2.5 m between 1876 and 1898 is believed to have occurred mainly between 1893 and 1898. The second was a rise of nearly 2 m in 1961.

---

[1] Schwabe (1844).   [2] Shaw (1926–32).   [3] Shaw (1933).

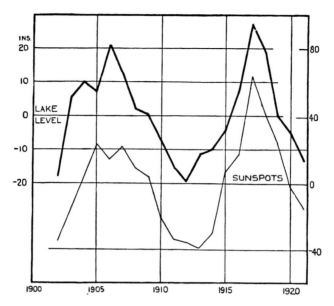

Fig. 1.2. The variation in the mean number of sunspots and level of Lake Victoria, East Africa, year by year from 1902 to 1921. (From Shaw, 1933.)

Failures like this gave weather cycles a bad name. In particular, attempts to demonstrate links between sunspots and the weather were frowned on by much of the meteorological establishment. This did not, however, prevent many determined souls labouring long and hard to provide better evidence of the existence of a link. By the late 1970s over a thousand papers had been published on the subject. But every new apparently convincing example of a solar–weather relationship was always subjected to searching statistical scrutiny by a sceptical meteorological community. Indeed, as late as 1978, a review paper by Barrie Pittock of the CSIRO in Australia, in *Reviews of Geophysics and Space Physics*, endorsed by a subsequent update in 1983 in the *Quarterly Journal of the Royal Meteorological Society*, summed up this scepticism.[4] He concluded that 'despite a massive literature on the subject, there is at present little or no convincing evidence of significant or practically useful correlations between sunspot cycles and the weather or climate'. Developments in the last ten years or so have produced results that have proved more difficult to dismiss so firmly. It is these developments that will be examined in detail later.

In part, these developments have been built on a nagging concern that it was difficult to dismiss some cyclic behaviour. The most obvious example is a tendency for many records to show a biennial oscillation. The

[4] Pittock (1978) and (1983).

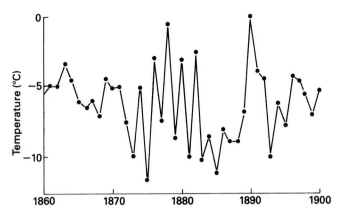

Fig. 1.3. The winter temperature record for Marengo, Illinois, showing that between 1873 and 1886 there was a marked biennial oscillation, but outside this period there was no such regular fluctuations.

example in Fig. 1.1 is echoed in many other observations. For instance, winter temperatures in the central United States in the 1870s and 1880s showed remarkably strong biennial behaviour for 11 years (see Fig. 1.3). But as with so many cycles, just when they look like a safe bet, they disappear only to re-emerge unexpectedly at some later date. Nonetheless, many meteorologists published papers noting the apparently impressive evidence of a biennial signal in many meteorological records.[5] By 1963 the weight of evidence was such that the climatologist Helmut Landsberg and colleagues[6] were able to conclude that there was 'no doubt the pulse, slightly in excess of 2 years in period, is a world wide phenomenon'. But they also described this phenomenon as a 'statistical will o' the wisp'. It is a measure of the problem that even now we do not have an adequate physical explanation of this biennial variability in surface weather data.

The search for the underlying cause of obvious roughly regular fluctuations has been more successful. These quasi-cycles may reflect fundamental properties of the natural variability of the global climate. As such they provide clues about the workings of the world's weather even if they may never amount to regular cycles. The expanding range of measurements of different aspects of the climate, such as upper atmosphere observations and satellite measurements, may hold the key to improved understanding. There have been two developing areas of analysis and expanding knowledge of longer-term global weather variability. First, from the early twentieth century onwards a series of studies had developed an orderly picture of

[5] An interesting history of these efforts is found in Chapter 4 of Labitzke & van Loon (1999).

[6] Landsberg *et al.* (1963).

large-scale oscillations in pressure patterns around the world. Later studies focused initially on pressure and sea surface temperatures across the tropical Pacific. By the late 1960s, a more comprehensive view had emerged of how events in the equatorial Pacific were linked to weather development at higher latitudes. These observations held the key to how quasi-periodic fluctuations in the tropics might lead to similar variations on a global scale. The close monitoring of subsequent El Niño events (see Section 5.4) in the tropical Pacific and in particular the record breaking events in 1982/83 and 1997/98 has since provided significant insights into how the atmosphere and the oceans interact to set up these approximately periodic oscillations. It also spurred on new interest in other regular fluctuations around the globe, some of which had been gathering dust since their discovery in the 1920s and 1930s. Now the study of 'oscillations' constitutes a growth area in meteorology.

The second major development in the cycles business was the discovery of an approximately regular reversal of the winds in the stratosphere over the equator. These measurements started in the early 1950s[7] and now clearly show the periodic behaviour of these winds, which reverse roughly every 27 months. This pattern has become known as the quasi-biennial oscillation (QBO). The importance of this upper atmosphere cycle is that, not only is it the most regular and predictable natural oscillation in the climate, but also it may be linked to the quasi-biennial feature in surface weather records noted above.[8] This development offers the prospect of being able to predict long-term variations at lower levels. Before this can happen there are two requirements: first, a satisfactory explanation of the periodic behaviour of the stratosphere as a whole, as the behaviour in equatorial region is pretty well understood, but links with higher latitudes are still the subject of debate; and second, a well-established physical link between changes in the upper atmosphere and consequent shifts at lower levels.

Alongside these advances in measurements has come improved understanding of the complexities of the global climate. In particular, the continued development of computer models of the climate has slowly

[7] See Labitzke & van Loon (1999) and Baldwin *et al.* (2001) for background on the discovery of the QBO and the current state of knowledge on this phenomenon.
[8] The acronym QBO is usually reserved for the stratospheric phenomenon, and its tropospheric cousin is sometimes termed the tropospheric biennial oscillation (TBO), but this gives the impression that the behaviour at lower levels is in some way more regular, so here we will use the generic term QBO for all quasi-biennial oscillations, recognising that the causes of this oscillation in the stratosphere and the troposphere may be different.

unravelled various aspects of the interconnectedness of all the components of the global weather system. But in spite of huge advances in computer power the models are still relatively crude and include greatly simplified assumptions to make the treatment of such parameters as cloudiness manageable. The central challenge is their handling of non-linear relationships between the various parameters in the model such as atmospheric pressure, temperature and wind speed. As with so many other areas of physics, the way round these problems is to establish that within certain limits there is a linear relationship between the various parameters. This means that for small shifts in the system the changes in one parameter are directly proportional to those in the other related parameters. This assumption that only the first-order terms are important and that higher-powered terms can be ignored makes the computation more manageable but imposes major limitations on the models.

The problem of handling non-linearity in physical systems has spawned a whole new area of science – chaos theory (see Section 8.1). This subject became highly fashionable in the late 1980s because of the new insights it provided and because it combined intriguing observations about the balance between order and disorder in the natural world with startlingly beautiful images of this balance. Its relevance here is that the theory had its origins in meteorology, and the weather arguably represents the ultimate challenge for the development of the theory. Whether chaos theory will play a central part in unravelling the specific issues surrounding weather cycles remains to be seen. What is apparent is that it provides a different way of looking at these issues and exerts a stern discipline on any attempts to provide any simple deterministic explanations for cyclic behaviour in the weather.

There is one other aspect of non-linearity that is frequently overlooked – the fundamental role of the annual cycle. It is central to the question of whether the climate is a chaotic system. Clearly, the atmosphere is a turbulent fluid and the chaotic behaviour of weather systems means that, in spite of the massive power of modern computers, numerical weather forecasts lose much of their skill beyond a week. But, although the atmosphere is chaotic on a day-to-day basis, the same need not apply to longer-term averaged conditions. We know that within the broad bounds of the annual cycle the climate in any particular part of the world generally sticks within prescribed limits: the temperature hardly ever rises above −20 °C at the South Pole, or falls below 20 °C in Singapore. The oceans interact with the atmosphere in a way that enables us to use knowledge of their slowly vary characteristics to make useful predictions of seasonal weather.

In addition, as we will see, the much longer variations associated with the ice ages can be largely explained in terms of changes in the Earth's orbital parameters. These results imply that some features of the climate are largely predictable.

There are, however, many examples of when the climate has behaved in a chaotic manner. At the end of the last ice age frequent sudden large changes in the climate occurred in a few years.[9] For instance, around 12 900 years ago, after a sustained warming, temperatures in the North Atlantic region plunged back to ice age severity. Events such as this have been linked with features of the collapse of the huge northern hemisphere ice sheets that was going on at the time. This suggests that, while there have been occasions when the climate behaved in a chaotic way in the past, for the most part its relative stability during the last 10 000 years or so, together with the regularity of seasons, suggest that currently it is not strongly chaotic. Nevertheless, it is not beyond the bounds of possibility that current global warming could eventually shift conditions into a more chaotic mode.

There is a more immediate potentially chaotic consequence of nonlinear behaviour in the climate. This is linked to the dominant role of the annual cycle and how it can combine with other forms of periodic or quasi-periodic fluctuations in the climate system to produce a variety of complex responses. In the case of interannual fluctuations like El Niño events, where the annual cycle is not a simple fraction of the longer-term fluctuation, the annual cycle can have a chaotic influence on the natural frquency of the longer periodicity. This could have profound consequences on our ability to forecast El Niño events.

Against this complicated background, this book will examine the evidence for weather cycles with two underlying aims. The first is to show that, whether or not the case for cycles stands up, the search for them sheds new light on how the climate works. The second is that, without a better understanding of the natural variability of the climate, it will be much more difficult to reach early conclusions on whether anthropogenic activities are having a significant impact. Tackling the threat of the build-up of greenhouse gases in the atmosphere as a result of the combustion of fossil fuels involves substantial adjustments in the nature of modern society. Although there has been considerable progress towards international action on this front, as the easy options are used up there will be a natural inclination to baulk at making more expensive and unpopular changes until the evidence of global warming is beyond doubt. But by then it may well be too late. So

---

[9] Taylor *et al.* (1993).

it is essential that we know more about how the climate can vary on its own accord to guide us in making these decisions.

With these thoughts in mind, we will explore in detail all the different aspects of the search for cycles and their physical explanation. But before we dive into the fascinating array of claimed cycles and proposed physical causes, the vexed issue of statistical analysis must be confronted.

Chapter 2

Statistical background

There are three kinds of lies: lies, damn lies and statistics.
Benjamin Disraeli (1804–81)

There is a fundamental presentational problem in discussing how to examine the evidence of cyclic behaviour in any stream of data recorded at regular intervals. This is that such examination usually requires some ferocious statistical analysis. Whether we are considering weather data or any other data recorded at set times (e.g. economic series), there is no way we can avoid this statistical approach. But to make it easier to present the underlying analytical techniques, the mathematics will be kept to a minimum in this chapter. This approach does, however, run the risk of glossing over the complexities of the analysis and giving the impression that the statistics can be put on one side. So to understand the problems of sifting through the evidence it is necessary to consider not only the description provided in this chapter but also the basic mathematics given in Appendix A. Failure to recognise the need for mathematical rigour can result in the reader being led up the garden path. It is important to belabour this unpalatable fact, as many of the published examples of 'weather cycles' have wittingly or unwittingly been the product of superficial or selective analysis of the available data.

Bearing in mind these words of warning, we must now consider the examination of the evidence of weather cycles. As explained in Chapter 1, the scale of the effort that has been devoted to the search for cycles is massive. So in addressing the results of this work and deciding what conclusions can be drawn from this effort, we have to consider first the nature of the data that has been collected and how it can be analysed. To do this we must start with the basic properties of time series.[1]

2.1 Time series

Any physical variable that is sampled at set constant time intervals can produce a time series. In the case of the weather, a series could consist of temperature or pressure measurements sampled continuously or every so many minutes or hours, or the amount of precipitation in successive equal time intervals. For the purposes of this book, which deals principally with cycles of periods longer than a year, we will be looking at series consisting of data that has been averaged over periods of a month or longer. So we will mostly be considering monthly or annual figures of average temperatures, pressure values or rainfall amounts for given locations or geographical areas.

The use of average figures in inevitable when we turn to indirect (proxy) data. Tree-ring widths, ice-core measurements, sedimentary records, cereal prices and wine harvest dates all by their nature will usually contain information about the integrated effect of a number of meteorological variables over a year or more. While it is possible to extract some seasonal information from variations of the form of, say, individual tree rings or the amount of snow making up a single annual layer in an ice core, and its properties, the amount of fine detail is inevitably limited. Given the emphasis on finding evidence of cycles of periods longer than a year, this is not a major drawback. It does, however, impose limitations on what can be extracted from the data, and it is essential that the effect of the form of the series is fully understood, otherwise there is a danger of falling into elementary traps.

The starting point for considering how much information can be extracted from recorded data is the fundamental property of time series. As a result of the work of the French mathematician Jean-Baptiste

---

[1] Much of this chapter consists of an attempt to encapsulate the essence of a variety of more detailed analyses of the statistics of time series and their spectra. So, rather than attempting to provide detailed references to these texts in this chapter the reader is advised to check the sources identified in the statistics section of the annotated bibliography. Only where recent new developments have been introduced into the study of time series will specific references be cited.

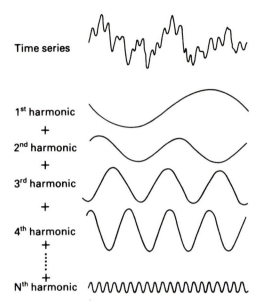

Fig. 2.1. A time series may be represented by the combination of a set of sine waves (harmonics) of differing amplitudes and phases.

Joseph Fourier, it can be shown that any time series can be expressed mathematically as the sum of a number of harmonics of differing amplitude (see Appendix A.3). All the statistical techniques that will be discussed here are designed to find out as much as possible about the harmonics that make up a time series. In principle, it is possible to compute the amplitudes of these harmonics for any series, and hence produce a complete picture of the harmonics present (Fig. 2.1). Until the advent of cheap powerful personal computers with standard programs capable of performing fast Fourier transforms (FFTs), these calculations involved a great deal of effort. For this reason, other more economical methods were often used to distil out the most important information. The ready availability of FFT programs does not, however, get away from the basic issue that it is essential to know what such computational aids are doing for the user; otherwise, there is a risk of attaching too much importance to what is generated by the computer. So the first thing to get straight is that the amount of information that can be obtained depends on three basic features: the sampling interval, the length of the series, and the accuracy of the observations. We will spend some time exploring the part played by these basic features of any series before going on to consider the relative merits of the various analytical techniques.

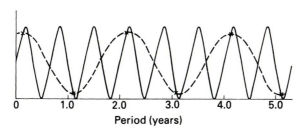

Fig. 2.2. The annual sampling of a series that consists of an 8-month periodicity can produce the misleading impression that there is a biennial oscillation.

2.2 Sampling

In any discussion of how the recording and presentation of meteorological data can influence the information that can be extracted from time series, it helps to take specific examples of what is normally recorded. If, for instance, we wish to examine the case for there being a periodicity of some number of years in winter temperatures in a given place, we must look closely at what is contained in the records. The standard form of temperature measurement is to record daily maxima and minima and average them to give the mean daily temperature. For longer-term studies, it is normal to work with the mean monthly temperatures. So the examination of the behaviour of winter temperatures over a long time for locations in the northern hemisphere will probably work with the mean temperature of December, January and February. This means that the series under scrutiny contains a single value for the winter temperature for each year, which is the average of some 90 pairs of daily readings, and no information about the rest of the year. This selective process has a number of consequences for the search for cycles.

  The first result of forming a single value for each year is that it effectively removes all information about the annual cycle in the weather. This is not only an inevitable consequence of working with annual figures, but is also essential as this annual cycle is by far and away the most dominant feature in all weather data and must not be allowed to interfere with other analyses. The other effect of working with annual figures is that it loses all sight of periodicities of less than a year.

  More important, if there do happen to be cycles with periods of, say, a few months, they could show up in the series in an odd way. For instance, an 8-month cycle could not be detected in its entirety, but could show up as an increase or decrease in temperatures in alternate years (Fig. 2.2), and hence would be interpreted as a 2-year cycle. Although the winter temperatures would indeed show a biennial oscillation, the diagnosis is inaccurate and

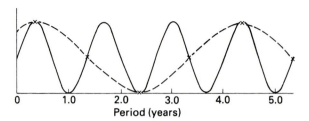

Fig. 2.3. The annual sampling of a series that consists of a 16-month periodicity will produce the misleading impression that there is a 4-year periodicity.

would lead to problems when seeking to attribute a physical cause to the observed behaviour. In effect, what has been detected is the different ('beat') between the two high frequencies, i.e. the 8-month cycle and the annual sampling cycle.

This result may not seem all that surprising; it is easy to see how the ups and downs in the shorter period cycle can modulate the annual figures. Somewhat less obvious is the effect of periods between 1 and 2 years. As can be seen from Fig. 2.3, the effect of sampling a periodicity of 16 months once a year produces a set of points that appear to be part of a four-year periodicity, i.e. a frequency of 0.25 cycles per annum (cpa). More generally, it can be shown that any frequency between 0.5 and 1.0 cpa will be reflected about the 0.5 cpa point and show up as a frequency between 0 and 0.5 cpa. This behaviour is familiar to statisticians and is known as 'aliasing', but can easily be overlooked when examining meteorological series. It is a fundamental feature of all sampling theory. It means that any spectral components with frequencies higher than the folding frequency (defined as the inverse of twice the sampling interval) will contaminate the spectral components below this frequency.

The way round aliasing problems is to work with a finer sampling interval. In the case of a standard temperature series the analysis could be conducted on the daily values. This approach does, however, have two drawbacks. First, it involves a great deal more data handling and computation, which is time consuming and expensive. Second, it gets in the way of the search for periodic behaviour at different times of the year, as there is evidence that periodicities for, say, winter and summer weather can be markedly different. So there may be good reason for picking out only parts of the total time series. Consequently, the dangers of aliasing must be addressed in any presentation of the resulting work. These issues are particularly important in the case where the meteorological variable is seasonally dependent or where the parameter, in the case of proxy data, reflects the weather for only part of the year. So examination of, say, drought indices for different parts of the world will reflect the rainfall amount during the growing season. Similarly, tree-ring widths will provide a measure of the conditions throughout the growing period while saying nothing about what happened during the dormant period. In such cases the problems of annual sampling cannot be got round by analysing data obtained at shorter time intervals, and must be accepted as a fundamental feature of the records.

The examples that have been cited so far may seem artificial and hence it may be helpful to consider a more realistic example of sampling problems. If, as Theophrastus proposed, there is some lunar influence on the weather (see Section 1.2) then this could cause some interesting effects when working with monthly averages. Because the lunar month is slightly shorter than the average length of a calendar month (29.53 days as opposed to 30.44 days) the analysis of a time series using monthly figures could produce a beat with a period of 33 months due to this small difference.

For the most part these statistical objections do not cause a great deal of difficulty because there is only limited evidence of any shorter-term cycles in the range of one month to one year (see Section 3.11) that interfere with the simple approach to searching for cycles in monthly and annual data (the one area that does need careful attention is considering shorter-term aspects of solar variability as the Sun has a rotation period of around 27 days (see Section 6.1)). Nevertheless, the important point to remember is that the sampling of time series imposes clear limits on what can be detected and the results can be misleading if they are not treated properly. So, throughout the book the consequences of the sampling interval in time series will be reiterated to ensure that their implications are not overlooked.

2.3   Length of record

The length of any time series has greater effects on the search for cycles. The first and most obvious is that there is no unambiguous information about periodicities longer than the length of the records. Although it is possible to draw some inferences about the longer-term components that could contribute to trend in the record (Fig. 2.4), this is limited by the accuracy with which the trend can be measured (see also Section 2.4). This means that in the case of a 100-year record, sampled annually, the useful information is restricted to cycles with periods from 2 to 100 years (i.e. frequencies from 0.50 to 0.01 cpa). Where the long-term trend represents an appreciable part of the variance, it is often decided to remove the trend before performing spectral analysis (see Section 2.6).

The second feature of the length of the record is that it limits the ability to resolve adjacent cyclic components when performing spectral analysis (see Section 2.6). While in practice the limit of resolution depends both on

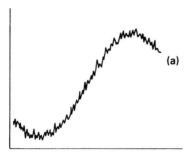

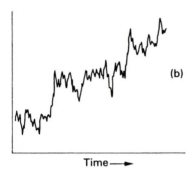

Fig. 2.4. (a) An example of a time series where it might be possible to make some useful inference about the existence of a cycle whose period is significantly longer than that of the time series; (b) a more typical example of a time series where all that can be estimated is the underlying trend.

the quality of the data and the mathematical techniques used to analyse the series (see Appendix A.3), there is a simple rule of thumb which can be used: it is not possible to resolve components that are closer than the reciprocal of the length of the record. So, with a 100-year record, it is not possible to separate two cyclic features that are less than 0.01 cpa apart. As will become apparent in later chapters, this theoretical limitation is of little practical importance, but it is a feature of the mathematical analysis which must not be overlooked, especially where attempts are being made to attach importance to the fine detail of power spectra.

2.4 Quality of data

Alongside the limitations placed on the analysis of time series by the sampling techniques and the length of the record are the problems of the nature and quality of the measurements. To examine these problems we must first define what the statistical analysis is trying to do. In essence this is a matter of detecting real periodic climatic signals in the presence of background noise. This definition is based on the assumption that the 'signal' can be attributed to specific physical causes, whereas the 'noise' is the unpredictable random component of the measurements. The noise arises from two sources. First, there is the obvious difficulty of the errors that can arise from the instruments used to make the measurements. Second, there is the background variability of the weather over time and space that has no coherence. These combine to produce either systematic or random errors, which can interfere with a statistical analysis that is aimed at identifying regular fluctuations (signal) and lead to confusion. It must always be remembered that even a series of random numbers can produce what look like 'significant' features (see Section 2.5) and that data containing principally random fluctuations will contain just this type of feature.

As far as the instrumental records are concerned (Chapter 3), most modern measurements are accurate and usually obtained under standardised conditions. So the chance of significant measurement errors being present in such data is small. But as we go back into the past the observations become increasingly suspect. Heroic efforts of scholarship have led to series being produced which remove the worst of these problems, but they can only go so far. In particular, systematic errors associated with a given site may be difficult to remove. For instance, depending on local topography and soil type, temperature measurements may prove to be more sensitive to certain weather situations: for example, night-time temperatures may be lower than in other similar areas. This amplification of specific weather types can produce significant distortions and misleading results. While corrections can be made, especially where a number of records can be used to form averages, care must be taken to check the provenance of the original data before too much credence is given to the statistical analysis of the fine details of the fluctuations over time.

These words of warning apply even more strongly where the meteorological variables show appreciable spatial variability. In particular, rainfall is notoriously variable from place to place. Where we have to rely on a record of measurements in one place, we must therefore pay particular attention to the limitations of the observations. The basic rule has to be that unless there is a sufficient spread of observations over a region to form an average value for rainfall, many of the fluctuations observed may simply reflect the local circumstances. In practice, there are few problems in using averages to cover significant geographical regions and smooth out the irregularities that occur in records for a single site. This is because, in general, if the periodicities are real they reflect some significant climatic effect that will inevitably affect a considerable area. But the reverse is not true.

Where attempts are made to extrapolate from a set of local observations to some wider conclusion much greater problems can arise. Even if the errors introduced by local conditions are randomly distributed, they can introduce apparently significant features (see Section 2.5 and Appendix A.9), especially in a relatively short time series. Where there are systematic errors, in particular if these vary appreciably over time (e.g. rainfall record which significantly under-recorded the amount of rain in the early years of the series), then this can lead to much greater difficulties. So whenever claims about cyclic behaviour in the weather are being made, it is important to know as much as possible about the underlying quality of the data before trying to attribute such fluctuations to real physical causes.

All these strictures apply with even greater force in the case of proxy measurements (see Chapter 4). The sampling interval is usually clear in the case of annual growth rings of, say, corals and tree rings. When it comes to establishing the rate at which ocean sediments have been laid down or stalagmites have built up, however, the challenges of defining an accurate timescale are much greater. Even where the timescale can be defined accurately, deciding whether non-meteorological factors have played a part in the rate of growth is central to the detective work. So, throughout the consideration of proxy data, there is a major requirement to keep a close eye continually on the quality of the data.

2.5   Smoothing: running means and filters

The simplest and most frequently used method of smoothing out a time series so that longer-term fluctuations can be identified is to form a running mean of data. In its most basic form this method consists of forming the average of a given number of successive points in the time series to produce a new series. Known as the 'unweighted' running mean, this approach is widely used and easy to apply. But it has a number of limitations, which need to be considered alongside the other methods of smoothing and filtering.[2]

To appreciate the impact of any smoothing operation on a time series we must consider how it affects the various harmonic components of the series. As we have already seen, any series can be represented by the sum of a set of harmonics. The easiest way to explain this is to take an example. If we are taking a 10-year unweighted running mean, the first obvious feature is that it will completely flatten out a 10-year periodicity of constant amplitude. This is because it will always be forming the average of one whole

[2] Burroughs (1978).

cycle wherever it starts from. Similarly it will remove all the higher harmonics that are an exact number of cycles in the 10-year averaging period (i.e. 5 years, 3.33 years, 2.5 years, etc.). It may also be apparent that its effect will be approximately to halve the amplitude of a 20-year periodicity, as the 10-year running mean will take the average of half this cycle as it moves along the series.

So far, so good, but when we come to look at what it does to some of the shorter periodicities, the problems start. Take, for instance, a cycle that has a periodicity of 6.33 years (i.e. it completes 1.5 cycles each 10 years). The 10-year running mean will thus form an average that contains the net effect of the additional half cycle as it moves through the series. Not only will this cycle be present in the smoothed series but it will also be inverted with respect to its original phase. As Appendix A.6 shows, after mathematical analysis of this worst case, 22% of this harmonic passes through the smoothing process and turns up as a spurious signal completely out of phase with the original harmonic in the unsmoothed series. This type of distortion, together with the presence of higher frequency features in different amounts, makes the use of the unweighted running mean both inefficient and potentially misleading.

To see how more efficient smoothing can be achieved, it is illuminating to consider the characteristics of an unweighted running mean in another way. The reason that high-frequency fluctuations get through is because of the way in which the smoothing deals with the data. Take, for instance, a time series of average winter temperatures that can fluctuate dramatically from year to year. These fluctuations may be random or contain some significant periodicities. The unweighted running mean is like a 'box-car' running through the series. Every data point within its span is given equal weight. So an extreme winter will enter the running mean with a sudden jump and exit in the same way, even though the running mean is designed to remove all such sudden changes. Given that we are only interested in the extremes to the extent that they are evidence of longer-term periodicities, it would be better if each data point came into the running mean gently, built up to a maximum in the middle and faded out again. Providing this approach both solves the problems of the unweighted running mean and does not introduce other distortions, it should be a better way of examining time series.

Appendix A.6 explores the mathematics of various weighted running means. The basic message of this work is that it is possible to design running means to act as relatively efficient 'low-pass' filters that remove virtually all the harmonics above a certain 'cut-off' frequency. The remaining harmonics are present in the series without any phase distortion, but close to the cut-off frequency their amplitude is substantially reduced. The choice of the mathematical form of the smoothing operation is a balance between achieving a sharp cut-off and minimising both the computational effort and the number of terms needed to produce the required smoothing effect. The latter is important because in general the sharper the cut-off the larger the number of terms that have to be used. This means that the ends of the series are effectively wasted in achieving an efficient smoothing, and if there are only a limited number of observations in the series this can be a high price to pay.

In practice, a reasonable compromise can be achieved by using the binomial weighting. Not only is this relatively effective in its frequency characteristics but it also has the benefit of arithmetic simplicity. It can be produced by one of two routes. First, and most direct, the desired level of smoothing can be chosen (e.g. 11-year running mean) and the appropriate coefficients (see Appendix A.6) applied to each successive set of adjacent terms in the original times series to produce the new smoothed series. Alternatively, the average of adjacent terms in the original series can be calculated and then the same operation performed on the new series, and so on until the required smoothing is achieved. The product of the first operation is the 2-year binomial running mean (identical to a 2-year unweighted running mean), the second operation produces the 3-year binomial running mean, the third the 4-year binomial running mean, and so on. The benefit of this approach is that not only is it arithmetically very simple, but also when displayed on a computer the effect of successive smoothing operations can provide some additional insight into how the variance is removed and hence some information about its frequency distribution.

Whatever approach is adopted has a basic drawback. This is because any low pass filter is not very discriminating. As a consequence, even purely random fluctuations when smoothed by running means in a relatively short time series can give the impression of there being significant quasi-cyclic fluctuations in the series, as Fig. 2.5 shows. This consequence of smoothing with a low-pass filter is known as the Schlutsky–Yule effect, after the two statisticians who demonstrated in 1927 that the nineteenth century 'trade cycles' could effectively be reproduced from a series of random numbers.[3]

If the main interest is the frequency distribution, it is possible to adopt a more selective procedure. The straightforward operation of smoothing time series using either a weighted or an unweighted running mean is only a specific example of the more general technique of filtering. Instead of simply working with a 'low-pass filter' which leaves the low-frequency

---

[3] Yule (1927).

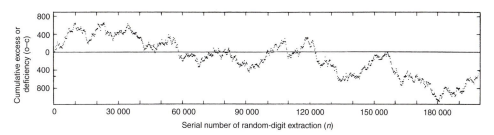

Fig. 2.5. An example of smoothing random numbers. (With permission of *Sky & Telescope* magazine.)

harmonics in the series unaltered and easier to see, there is no reason why this practice should not be extended to suppress both high- and low-frequency components and let only a limited range of frequencies through. The advantage of this process is that, unlike harmonic and spectral analysis, which we will come to in Section 2.6, it permits the examination of the persistence of periodic features throughout the duration of the series. By comparison, the power spectrum is only about the mean amplitude of apparently significant oscillations while their variation over time is transformed into other components of the spectrum. So if the amplitude of the periodicity changes appreciably over time, spectral analysis may give a misleading impression of the nature of fluctuations.

This distinction is important. In later chapters, it will become apparent that convincing evidence of periodic behaviour can come and go with tantalising regularity. After several periods a cycle can suddenly disappear, only to reappear at some unspecified interval later, or shift phase and amplitude, or disappear for good. So mathematical techniques, which expose all these frustrating differences, can help to pin down the physical reality of causes of any supposed cyclic behaviour.

Ideally, a filter should pass all frequencies within a narrow band without any change in amplitude and completely suppress all other frequencies (Fig. 2.6). In practice, this is impossible to achieve and compromises have to be made in choosing a filter which provides the best combination of removing unwanted frequencies and leaving largely unaltered the frequencies of interest. The underlying approach to narrow band filtering is explained in Appendix A.6. What this demonstrates is that the general form of such a filter is an oscillation of the required frequency, with the amplitude increasing from a small value up to a maximum and then reducing again. The number of points used in the filter defines the bandwidth of the filter and hence the number of oscillations included in the computation – the greater the number of points used in the filter, the narrower its bandwidth. But, as with all smoothing and filtering operations, there is a pay-off between the narrow

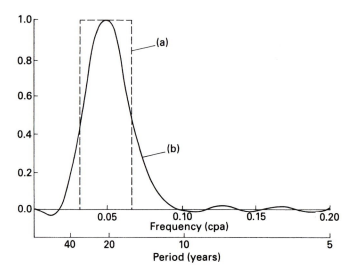

Fig. 2.6. A comparison between (a) an ideal statistical filter, which removes all the unwanted periodicities and leaves unaltered those periodicities which are of interest, and (b) what can be achieved in practice. In (a) the filter transmits periods between 15 and 30 years whereas (b) only transmits 20.6 years unaltered and reduces the amplitude of all other periodicities.

bandwidth of the filter and both the computational effort and the available data. In particular, if a sharp filter involves using a significant number of the available data points to compute a single point in the smoothed series, it limits the scope of the analysis. Moreover, if the data contain a considerable amount of noise, too precise a focus on a narrow frequency range may serve little purpose. So, as with so many aspects of the search for weather cycles, there is a compromise to be struck in dealing with the limitations of the data.

2.6 Harmonic analysis and power spectra

The underlying principle of harmonic and spectral analysis of time series is the work of the nineteenth-century French mathematician, Fourier. As noted in Section 2.1, he demonstrated that a function could be expressed as the sum of a set of harmonics of differing amplitudes (see Appendix A.3). This means that there is a direct mathematical link between a given time series and the function which describes the amplitude distribution of its component harmonics. So it is possible to calculate the spectrum of the harmonic function that will be formed by combining any set of harmonics of specified amplitudes. Because of the complementary nature of these mathematical operations the function is usually described as being transformed into its harmonic spectrum and vice versa. In recognition of this work these operations are usually known as Fourier transforms.

Over the years, a range of techniques has been developed to examine the harmonic components of time series. Many of these were designed to tackle the problem that a complete analysis of all the possible harmonic components involved a great deal of arduous arithmetic. For while the fundamental mathematics of this form of analysis has been understood for over 150 years, the practical application of this knowledge to lengthy time series was limited by shortage of computational power and efficient algorithms for the rapid calculations of Fourier transforms. Since about 1970, the ready availability of powerful computers and efficient mathematical programs has meant that the harmonic analysis of large amounts of data has been a practical proposition.

The advances in computer technology mean that the elegant mathematic techniques that earlier workers used to minimise the effort in identifying the existence of the important harmonics in time series (e.g. correlograms and filter analysis) have been made largely redundant by the brute force approach of modern computers. While computers have provided the luxury of being able to extract all the information in available time series, they have tended to make it more difficult for researchers to exercise a critical approach when presenting their results. All too often power spectra are produced which contain a whole range of features, many of which are identified as being statistically significant. This means that someone new to the subject may find it difficult to discriminate between the various features and establish what are the physically significant results. Since this sifting process is central to the theme of this book, it is best to concentrate on what is involved in modern spectral analysis of time series and how increasingly sophisticated techniques have been used to squeeze the maximum out of the data. In so doing, the aim will be to steer between the opposite extremes of being taken in by the results of computer wizardry and of dismissing anything which is not overwhelmingly cyclic in origin.

The ready availability of computer programs which can rapidly calculate the Fourier transform of lengthy time series means that we need to focus on the information in the spectra that are produced by this process. To do this it is perhaps easiest to consider some examples of time series and their complementary spectrum. This pictorial approach is backed up by the basic mathematics in Appendix A.3. Starting with the most trivial example, the monthly temperature record for a mid-latitude site in the northern hemisphere would be dominated by the annual cycle. So if we computed the Fourier transform of a number of years' observations, the spectrum would be dominated by the annual cycle (Fig. 2.7). In practice, what would normally be computed is the transform of the *variance* of the temperature

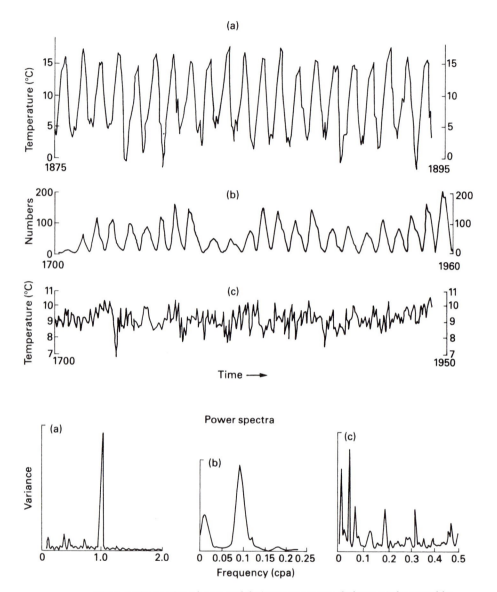

Fig. 2.7. Time series (above) and their power spectra (below): (a) the monthly temperature record for central England between January 1875 and December 1895, (b) the number of sunspots during the period 1700 to 1960, and (c) the annual temperature for central England during the period 1700 to 1950, showing how with increasing irregularity in the time series the power spectrum becomes more complicated.

observation from the annual mean. This is defined as the square of deviation of any given observation from the mean. Then the transform of the time series of the variance is computed to produce the *power spectrum*[4] and this is a direct measure of amount of the variance that is due to each harmonic in the spectrum. In the case of the monthly temperature record in Fig. 2.7(a), virtually all the variance would be found in the annual cycle with a residual scattering of other lesser components that reflect all the other fluctuations from month to month and year to year.

A slightly less trivial example of a Fourier transform can be found in sunspot numbers. Given that so much of the search for cycles in the weather has been associated with finding links with solar activity, it is a good example to consider. As Section 6.1 describes, sunspot numbers show pronounced cyclic behaviour with the major fluctuation having an approximately 11-year period. In addition, successive 11-year peaks show a periodic variation in intensity that reflects a periodicity of around 90 years. So the power spectrum obtained by calculating the Fourier transform of a lengthy series of sunspot numbers shows two pronounced peaks (Fig. 2.7b). Because the cyclic variations are not precisely 11 and 90 years, the power spectrum shows relatively broad peaks that reflect the varying period from cycle to cycle. But the important feature is that the power spectrum confirms what is evident from inspecting the record of sunspot numbers – almost all the variance in the last 200 years or so can be attributed to the 11-year and 90-year periodicities. As in the case of the annual temperature cycle, the link between the time series and the power spectrum is relatively easy to see.

This direct link becomes much less obvious in the case of a typical meteorological record where we are interested in identifying periodicities in the range 2 to 100 years. Here there is no obvious cyclic behaviour in the variances of the year-to-year figures from the long-term mean. So the power spectrum (Fig. 2.7c) will contain a number of features of varying magnitude. But deciding which of these features is both statistically significant and of physical significance requires careful analysis, and will be considered in more detail in the next section. For the moment, the important fact is that by calculating the Fourier transform of a time series, it is possible to produce the power spectrum of all the harmonics that uniquely define the observed series. Conversely, if we knew only the power spectrum it would be possible to recreate the time series by the reverse calculation. This complementary

---

[4] This nomenclature reflects that used in electrical engineering where the power associated with an alternating current is proportional to the square of the amplitude of the current. So, given that the statistical definition of variance is in terms of the square of the deviation from the mean, it is standard practice to define the transform as the power spectrum.

nature of the time series and its power spectrum is not only an expression of the mathematical link between the two: if the observed power spectrum were a measure of the physical behaviour of the weather in the future as well as in the past, then it could also be used to forecast the weather. Successful forecasting is the true test of reality of the supposed cyclic behaviour of the weather.

The development of increasingly powerful computers and programming methods has led to considerable efforts to squeeze more information out of time series than is available from basic Fourier transform methods. In particular, where the series is of only limited duration there is potential benefit in using the available data to the fullest degree. The main problem to be addressed is that in normal circumstances the transform of a time series assume that there is no information available outside the span of the record. This means that in effect the variance observed within the records drops suddenly to zero at the ends of the series. This sharp truncation produces difficulties with the spectral analysis that are analogous to the effects produced in using the unweighted running mean (see Section 2.5). The way round this problem is similar to that used in forming weighted running means, in that the discontinuity at the beginning and end of the series is removed by giving less weight to the ends of the record. But this has the disadvantages of discarding some real information and reducing the resolution of the spectral analysis.

Another way round the problem is to make some assumptions about the nature of the time series outside the span covered by observations. The best-known example of this approach is called maximum entropy spectral analysis (MESA), and this method became popular in the 1970s.[5] The nub of this technique is to extend the record without adding or taking away information. The principle which enables this to be done draws on the probability of the harmonics having certain amplitudes based on the available information, but is maximally non-committal with regard to the unavailable information. Known as the principle of maximum entropy, the approach can be used effectively to generate additional length to any time series so that the fullest use can be made of the information in the original observations. But it must be used with care, because while it produces improved resolution and some extra information in the form of smoothing of the power spectrum, it cannot produce more information that was present in the initial time series. For example, Fig. 2.8 shows the consequences of applying MESA techniques to annual rainfall figures for Kew between 1697 and 1975.[6] The effect of increasing application of MESA is to throw up more

[5] Ulrych & Bishop (1975).   [6] Tabony (1979).

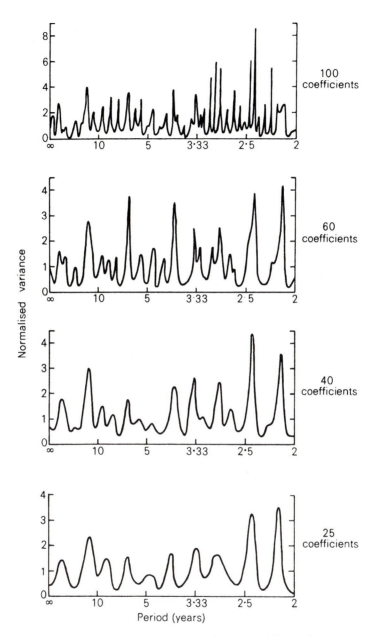

Fig. 2.8. Maximum entropy power spectra for the rainfall record at Kew, England, for the period 1697 to 1975. The increased application of the MESA technique to a time series where much of the variance is random produces an increasingly noisy spectrum with little or no additional useful information. (From Tabony, 1979. With permission of the Controller of Her Majesty's Stationery Office.)

and more sharp peaks. But as will be seen in Section 3.3, these peaks are not reproduced in other parts of the rainfall records, and may be nothing more than increasingly detailed resolution of the noise present in the time series.

This example exposes the fundamental limitation of MESA. In essence its success depends on the signal-to-noise ratio in the time series being high and has worked to good effect in certain areas, such as geophysics.[7] In the right circumstances, when it can be assumed that certain period-icities are present in the series, it is possible to optimise the available information about these periodicities. This is of particular value when the series is short compared with the periodicities that form its princi-pal components. But where the series is principally noise, the technique has real limitations. At best it will only enable the researcher to explore the noise in ever more excruciating detail. At worst there is a danger of prejudging the outcome of the analysis with misleading conclusions. As will become obvious in later chapters, in only rare instances do meteoro-logical series meet the signal-to-noise criteria that can exploit the advan-tages of MESA. Furthermore, as Fig. 1.2 demonstrated, time series which look eminently suitable for MESA can prove to be a snare and delusion in subsequent years. Clearly this is a technique that needs to be used with circumspection.

There is one other aspect of spectral analysis that needs to be con-sidered. This is the effect of long-term trends in the data. As explained in Section 2.3, the length of the time series limits the information about long-term variations. So if the available record shows a significant trend, which at first approximation is linear, spectral analysis will not provide a reliable analysis of the longer-term nature of this trend. But the mathe-matical process by which the harmonic components in the time series are derived produces an odd result. Because the analysis assumes that there is no variance outside the range of the observations, the trend is viewed as a triangular ramp (Fig. 2.9). The combination of harmonics of different amplitudes that make up this linear function will have a significant impact on the analysis, as their amplitude is inversely proportional to frequency. Since the power spectrum is made up of the square of the amplitude of the harmonics, the linear trend will be transformed into a contribution to the power spectrum, which is inversely proportional to the square of the frequency. So, whether or not the trend is real, it will have a major im-pact at longer periodicities. Since this may cause problems in analysing the other components of the time series, it is better if the trend is removed.

[7] Bath (1974).

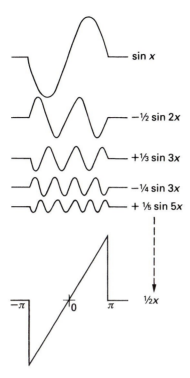

Fig. 2.9. A linear trend over a period in a time series is represented by a set of harmonics whose amplitude is inversely proportional to the frequency.

If this is done, the series is said to be 'detrended', or 'prewhitened' (see Appendix A.10).

This brief set of observations aims to introduce the complexities that underlie the spectral analysis of time series. The basic mathematics is described in Appendix A.3, A.4, A.5 and A.8, but even this more detailed presentation can only scratch the surface of the arcane statistical procedures that are available to analyse the harmonic components of time series. For the purposes of this book, however, it is necessary to have a feel for the limits of the analysis but it is not essential to have a complete understanding of the analytical techniques. What is important is to know what the statistical techniques can and cannot do, so that claimed features are subjected to proper critical scrutiny. But before we consider the meteorological evidence there is one further aspect of time series that exerts a major influence on the search for cycles. This is the nature of the random errors in the data. This represents the principal challenge that the statistical techniques described here are designed to overcome. It has been touched on in Section 2.4 but we now need to consider its implications for spectral analysis.

2.7   Red, white and pink noise

The effect of errors in observations and the natural variability of the weather exert a powerful influence when a meteorological time series is transformed into a power spectrum. The basic question is, what spectrum would be obtained if there were no real cycles present and if all we were looking at was observational and meteorological noise.

Before deciding what such noise will look like, we must be clear what we are talking about. As Section 2.4 explains, and Chapter 3 will explore in more detail, there are many sources of systematic error that can lead to short-term changes or long-term drift in time series. These will cause problems, but with careful research these can be largely eliminated. Of greater concern are the random errors implicit in the observations, which cannot be removed by better examination of the original data. As a general rule, if these errors are truly random, it can be assumed that they occur on every timescale and so are capable of contributing to every possible frequency with equal probability. The spectrum of such random fluctuations will be a constant amplitude for all frequencies. What this means is that in any unit frequency bandwidth the probability is that there will be an equal amount of variance. The standard term for such a frequency distribution is 'white noise' – this expression is derived from optical spectroscopy where 'white' light contains all the visible frequencies.

The effect of the presence of random errors on a power spectrum containing real features is in theory to add in a uniform background of noise. But, in practice, we are analysing only a limited time series and with it a limited selection of random errors. As a consequence, the spectrum of errors may appear to contain real features (see Section 2.5 and Appendix A.9). So it is necessary to calculate the probability of these random features exceeding certain levels. It is then possible to estimate the chances of a given peak being the product of random fluctuations or a real feature.

This picture is complicated by the natural variability of the weather that reflects all the complex interactions in the global climate. As will be seen in Chapter 5, there are a number of components of the weather machine that tend to damp out the more rapid fluctuations. Slowly varying factors like snow cover, polar ice, sea surface temperatures and soil moisture build in inertia and mean that the weather has a 'memory' and so is more likely to exhibit greater low-frequency fluctuations than higher-frequency ones. Again in the terminology of optical spectroscopists, such noise is defined as being 'red', denoting that its distribution is weighted towards lower frequencies. In effect, because the weather has a better recollection of recent events, the short-term variations are damped out more than the longer-term ones, because with the passage of time the connections become more tenuous. The theoretical distribution of red noise depends on the assumptions made about how any connections between successive events decay over time (see Appendix A.9). The important point is, however, that an estimate of the likely distribution of red noise can be made and this has to be used to assess the significance of what appear to be real features. In practice, this means that lower-frequency/longer-period cycles have to contain a greater proportion of the observed variance to achieve the same significance as higher-frequency/shorter-period features.

These observations about the frequency distribution of noise may be surprising. The natural expectation is that short-term fluctuations in the weather are greater and more rapid. But to find the frequency distribution of noise, we need to know the amount of variance in a unit frequency bandwidth. It is important to note that in this book the frequency and its reciprocal, the period, of any cycle are used interchangeably. We will concentrate principally on the frequencies in the range 0.005 cpa to 0.5 cpa (periodicities from 200 to 2 years). This is a very narrow frequency spread when compared with fluctuations on the scale 1 week to 1 day that cover the range 52 to 365 cpa. So, although short-term fluctuations in the weather may appear dramatic, when spread over this much greater frequency range their contribution to the power spectrum for unit frequency interval is less than the longer-term fluctuations.

Normally, the errors in instrumental records will be dominated by the variability of the weather, so that red noise is the best approximation. But in the case of proxy data the situation is more complicated. Because the link between the observed variable (e.g. tree-ring width) and the meteorological parameters (e.g. rainfall) is the subject of considerable uncertainty, there are bound to be random errors. While these errors can be reduced by the careful calibration of more recent proxy data using modern meteorological records, the problem cannot be eliminated. As a consequence, there will be considerable random error in the inferred meteorological variability. This will produce white noise in any computed power spectrum. At the same time, the underlying weather will have contained red noise. Moreover, in some cases such as tree rings there are additional reasons for red noise. Because tree growth depends on groundwater reserves, the ring thickness depends on rainfall not only in the current year, but also in the previous year or longer. So spectral analysis of proxy data will contain both types of noise; this combination is often referred to as 'pink' noise. This means that in any consideration of the significance of spectral features obtained from the analysis of proxy data an estimate of the pinkness of the background noise must be calculated.

2.8   Wavelet and singular spectrum analysis

Wavelet analysis is a form of spectral analysis that has become exceedingly popular in recent years. In effect, it combines features of various of the methodologies described above. It is designed to examine how the power spectrum of a time series varies over the time of the record.[8] It has been the subject of considerable debate as it was seen by some scientists as being nothing more than a colourful way of presenting complicated data. Because the technique involves a transform of a one-dimensional time series (or frequency spectrum) to a two-dimensional time–frequency image there is bound to be the appearance of additional diffusivity. In the realm of weather cycles, however, the presentational device outweighs any additional imprecision as it highlights the natural variability of the climate system by mapping these changing patterns in a statistically significant manner. It also addresses the fundamental limitation of the Fourier transform of a time series, which contains harmonics whose amplitude varies significantly throughout the length of the record, such that it effectively 'smears' these variations out in calculating the spectrum for the whole series.

Wavelet analysis confronts the challenge that conducting a running Fourier transform of the available time series simply by using a certain window size and sliding it along in time has serious limitations (see Appendix A.7). This direct approach would give us information about the frequency spectrum, but treats different frequencies inconsistently. Whatever window width is chosen, it would be too small to resolve different low-frequency oscillations. Conversely, at high frequencies, although the resolution is fine it would be better to have a narrower window to examine the shorter-term time variations.

What wavelet analysis does is to combine both a weighted window and a defined number of oscillations of a given frequency within this window. This is an analogue of the weighted filters described in Section 2.5, with the weighting of the window chosen to avoid the pitfalls of using a simple unweighted 'box-car' form. The most obvious choice for studying weather cycles is to use a Gaussian envelope, as the Fourier transform of a Gaussian is another Gaussian. The wavelet formed by a cosine wave and a

---

[8] Torrence & Compo (1998).

Gaussian envelope is called the 'Morlet' wavelet (see Appendix A.7, Fig. A.7). The benefit of this type of analysis in looking at the behaviour of climate systems is that it provides a clear representation of how the variance at different frequencies changes throughout the duration of time series under investigation. This has the particular advantage of showing clearly how different periodicities come and go throughout the record (e.g. Fig. 3.8). This transient behaviour is a common feature of many climatic 'oscillations' that will frequently be the subject of detailed discussion in this book.

Other ways of extracting information about temporal variations of power within time series include *singular spectrum analysis*. This adopts a different approach to manipulating the information in time series (see Appendix A.8). The advantage of this technique is that it identifies recurrent time patterns in the single time series, which is particularly helpful in isolating anharmonic oscillations with fluctuating amplitudes from noisy data.[9]

[9] Venegas (2001).

Chapter 3

Instrumental records

The Great Tragedy of Science – the slaying of a beautiful
hypothesis by an ugly fact.

T. H. Huxley (1825–95)

Armed with a knowledge of the techniques for analysing time series, we
must now get down to the business of examining examples of the efforts
that have been made by meteorologists to produce evidence of cycles. The
best place to start is with instrumental records that enable us to consider
the behaviour of a single variable (e.g. temperature, rainfall or atmospheric
pressure) as a function of time. The aim will be to sift through the evidence
and draw up a balanced assessment of the case for cycles of any given
period. We will review a cross-section of the work that has been done to see
which periodicities show up most frequently and which seem to be peculiar
to a given record. From this inventory we will then be in a position to move
on first to the wide range of data, which contain indirect information about
the weather (proxy data), and then on to possible physical explanations of
what might have caused such regular fluctuations.

As has been made clear in Chapter 1, there has been a huge amount
of work done over many years to demonstrate cyclic behaviour in weather
records. For a number of reasons we will pass over much of this work and
concentrate on more recent work. First, many of the early studies have been
the subject of critical review and have been discounted by many meteorologists. More damning, as noted in Chapter 1, it seemed that as soon as
evidence of many of the most convincing cycles was published, they ceased
to be a feature of the weather. Second, the data used often suffered from
a variety of limitations that constrained the value of the resultant analysis.
Third, the advent of computers and a wide variety of readily available

statistical programs has meant that more extensive and thorough analysis of the data is now possible.

The considerable work that has been done recently to improve the quality of the series based on early observations, together with the expanding data base obtained in recent decades, provides a better starting point for the statistical analysis. Combined with the additional computing power at the disposal of meteorologists, this effort has produced much more detailed analysis of possible cycles. This means that the most sophisticated and thorough studies on the most comprehensive databases have been done in recent years. Before we look at the work there is one important word of warning. Because the analysis looks so much more impressive, it is easy to be lured into believing that there is more to the results than is really the case. The combination of beautifully smoothed power spectra and a plethora of significance figures can be disarming. The real tests of claimed cycles are twofold. First, how often do they occur in independent sets of data? Second, and more important, as we shall see when we come to Chapter 7, can they be explained in terms of a causal physical mechanism?

There is one other area of data management that should be mentioned at the start. This is the major work in recent years at the European Centre for Medium Range Forecasts (ECMWF) and the US National Centers of Environmental Prediction (NCEP) and the National Center of Atmospheric Research (NCAR) to reanalyse all the data used for weather forecasts since the 1950s.[1] So far, the published data from the ECMWF only covers 15 years (1979 to 1993), but work is in progress to extend the analysis back to the period from mid-1957 to 2001. The NCEP/NCAR published global results already extend back to the 1958 and to 1948 for the northern hemisphere. These reanalyses are providing a much more comprehensive picture of the climate. Exploiting the modern computer models that are used in numerical weather forecasting work to assimilate all the observations that were collected as part of the standard weather forecasting operations of the time, it has been possible to produce a standardised presentation of day-by-day changes of the global climate. Because of limitations in the models and changes in how the data were collected, this work needs to be treated with great care in detecting climatic trends, but it is invaluable in looking for periodic changes in global patterns weather patterns.

[1] An introduction to the NCEP/NCAR project is given in Kalnay *et al.* (1996), and the website address is: http://www.cdc.noaa.gov/cdc/reanalysis/reanalysis.shtml. An introduction to the ECMWF reanalysis work can be found on: http://www.ecmwf.int/era/index.html.

3.1 Central England temperature record

A series of monthly temperatures prepared by the late Prof. Gordon Manley for lowland central England is the longest homogenous temperature record in the world. Extending back to 1659, it not only provides an excellent series in which to explore the evidence of cycles in temperature records, but also demonstrates the effort that is involved in producing such series.[2] It was the product of many years of thorough and diligent scholarship by Prof. Manley, and involved a number of interlinked efforts. The first task was to search out and bring together all the records that had been accumulated by a bewilderingly diverse array of amateur observers before the days of official meteorology. The gathering together of the records was only the start of the analytical problems. First, there was the question of how the measurements were made. Back to the early nineteenth century the combination of reasonable standard observations plus sufficient numbers of overlapping records enabled useful checks to be made of the reliability of the observations and adjustments made for, say, measurements at different times of the day. But earlier records posed greater problems. Before 1760, some of the best-kept records depended on having thermometers exposed in well-ventilated north-facing fireless rooms. A further complication was that prior to 1752 it was not possible to obtain monthly means capable of comparison with those of England today; neither could they be compared with those of contemporary western Europe unless there were daily observations to cope with the change from the Julian to the Gregorian calendar – not adopted in England until 1752, by which time the difference amounted to 11 days.

One way in which the inconsistencies between different sets of observations could be produced was by using weather diaries. These provided confirmation of relevant weather events such as snowfall and days with frost, which help to build up a better picture. But even when the instrumental discrepancies had been ironed out there were real differences that required careful treatment. As Manley noted, there were broadly six types of inland site. These were urban, favoured well-drained slopes, hilltop, lakeside normal open lowland, frost hollows, and exceptionally sandy soils. The physical differences between these different sites had to be assessed to enable him to produce the most probable mean temperature for central England. In particular, the effects of urbanisation and slight changes

---

[2] Manley (1974) and Parker, Legg & Folland (1992). This monthly series is now routinely updated by the Hadley Centre and available on its website: http://www.met-office.gov.uk/research/hadleycentre/obsdata/index.html.

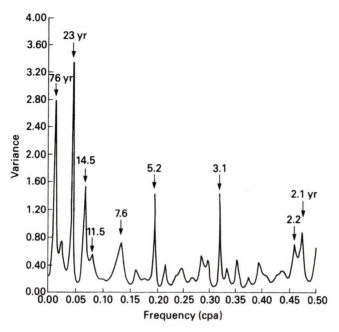

Fig. 3.1. The power spectrum of the detrended central England temperature record for the period 1700 to 1950 showing notable periodicities at 2.1, 2.2, 3.1, 5.2, 7.6, 14.5, 23 and 76 years. (From Mason, 1976.)

in the siting of the instruments had to be identified and corrected for, otherwise the long-term fluctuations and trends, which could be a product of these local changes, might be erroneously interpreted as real climatic shifts. If these effects had not been removed, the statistical tests could have produced misleading features, especially in the case of calculating power spectra (see Appendix A.9).

Manley's efforts have, however, produced a series that provides a particularly useful source for investigating the evidence of cycles in temperatures in England on a timescale from 2 to 200 years. An analysis by the UK Meteorological Office of this series,[3] working with the data from 1700 to 1950 and using the MESA method (see Section 2.6 and Appendix A.5) produced the power spectrum shown in Fig. 3.1. This analysis first removed the linear long-term trend that contained about 20% of the variance. The significant features in the spectrum are at periods of 2.1, 2.2, 3.1, 5.2, 7.6, 14.5, 23 and 76 years. The first peak, which may be related to the quasi-biennial oscillation (QBO) that was introduced in Chapter 1 and will reappear at regular intervals throughout this book (see in particular Section 3.9), contains about 10% of the total variance and is significant at the 5% level. Some of

[3] Mason (1976).

the lower-frequency peaks may well be the result of non-linear interactions between higher frequency periodicities, but the 23-year peak, containing about 8% of the variance and significant at the 0.1% level, may be associated with the double sunspot cycles (see Section 6.1). The spectrum is also interesting because certain features are absent. In particular, there is only a very weak feature around 11.5 years, whereas on the basis of many of the other results we will be discussing we might expect to find a strong peak linked with the sunspot cycle. Also there is no evidence whatsoever of the 18.6-year lunar cycle (see Section 6.2) which features so frequently in later discussions.

More recently, the series has been reworked using wavelet analysis and singular spectrum analysis (SSA).[4] In both studies, the periodicities at around 7, 14 and 24 years are detected. In addition, the 5-year feature is evident in the wavelet analysis, as well as a previously unreported feature at 102 years. More important, this work shows that the interdecadal and century-scale variability is strongly dependent on the period of the analysis. When the analysis is conducted for the two periods 1659–1856 and 1856–1990 only the 7-year and, to some extent, the 14-year peaks appear in both spectra. This changing behaviour over time is a recurrent feature of many meteorological time series and is something we will have to grapple with throughout this book.

An analysis supported by the German Research Ministry went into more detail by examining the evidence of cycles in the data for each month, as well as annually, over the period 1660 to 1977.[5] This produced not only much more detail about possible cycles, but also considerable confusion as to what was and was not physically important. The variance spectrum of the annual figures was broadly similar to that produced by the other analyses. There were, however, interesting differences which show the problems of using slightly different lengths of record and mathematical techniques. At the low-frequency end, the significant features are at 25 and 100 years as opposed to 23 and 76 year. These differences are, in fact, within the statistical uncertainties of the techniques used to produce the spectra. But, given that so often the link with other physical processes depends on the coincidence between observed periodicities, it is important to highlight the differences that can occur using approximately the same series and the same form of analysis.

The discrepancies between the two analyses of the annual figures are, however, small compared with the differences between the variance

---

[4] See Plaut, Ghil & Vautard (1995) for the SSA analysis and Baliunas *et al.* (1997) for the wavelet analysis.

[5] Schonweise (1980).

spectra for each month. The 100-year cycle is only significant in August, September, October and December, while there is a strong 200-year signal in January, and less so in February and March. These cycles are absent in other months, although April, June and especially November have a 67-year cycle. The annual 25-year cycle is made up of a variety of monthly features of varying significance from 22 to 33 years, between March and August, but which are absent in other months. Between 9 and 15 years, there are a few features of low significance but no consistent picture. Similarly, there is an accumulation of features around 5 years, but these are scattered over quite a frequency range, are mostly of low significance and are missing in several monthly records. Most months feature a peak between 2.9 and 3.9 years, but again they are spread over quite a frequency range. The most consistent picture is in the 2.1- to 2.8-year range where every month has at least one peak. In particular, 2.2 to 2.4 years appears frequently and is often highly significant. This appears to be evidence of the tropospheric version of the QBO.

This first example of the search for cycles in well-established and lengthy instrumental records immediately throws up the problem that will dog our search. This is that there is no shortage of 'significant' cycles. What is missing is evidence of the same periodicity showing up at all times of the year and throughout the entire record. Apart from the feature that may be related to the QBO, everything else come and goes in a tantalising way. So the basic question to be addressed as we go from study to study is whether the evidence adds up in favour of certain periodicities, or whether we are merely accumulating more and more cycles of different frequencies. If it is the former, we have something that is worth seeking a physical explanation for; if it is the latter then the exercise is no more than cataloguing the almost infinite natural variability of climate.

3.2 Other temperature series

Given the large number of lengthy temperature records that exist to be drawn upon, especially for European sites, surprisingly little work has been published on the evidence of cycles in such series. When compared with the amount of work that has been done on other records, such as rainfall and atmospheric pressure, this suggests that the absence of published material may be evidence of a lack of success in producing significant results rather than a failure to do the work. This negative conclusion is supported by an early study of a series representing Central European temperatures constructed from observations in Vienna, Berlin, Paris and

the Netherlands for the period 1761 to 1960, which detected no significant periodicities apart from the ubiquitous tropospheric quasi-biennial
oscillation.[6]

A study of January temperatures for 12 stations in the eastern United
States and Canada had a little more success.[7] This examined a composite record from these stations stretching from South Carolina to New
Brunswick and covering the period from 1975 back to the late nineteenth
century in most cases and 1779 in one case. Along with the usual evidence
of some marked periodicity at about 2.2 and 2.5 years, there were significant spectral peaks at 4.5, 9 and 20 years. This work concentrated on the
evidence of the 20-year cycle and examined its variation over time using filtering techniques. The interesting feature of the study was that the
20-year cycle was most prevalent and pronounced in all the records during the period 1920 to 1960. The investigators suggested that the behaviour
could be linked with the fact that this period was marked by stronger than
normal circulation in mid-latitudes of the northern hemisphere. This, combined with the fact that zonal circulation in this region peaks each year in
January, suggests that the evidence of periodicity may be dependent on
wider climatic shifts. As we will see, this is an argument that surfaces from
time to time to explain the transient nature of what for a while looks like
convincing cyclic behaviour.

A 22-year cycle turns up in one other important temperature series.
This is in the global record of marine air temperatures, consisting of ship-
board temperatures made at night. It is one of the most reliable measures of global temperature trends. Work by Nicholas Newell, a scientist in
Arlington, Massachusetts, and colleagues at the Massachusetts Institute
of Technology analysed the fluctuations after filtering out the long-term
variations associated with a major dip around 1910 and the warming trend
of the last 70 years.[8] Over the period 1856 to 1986 the power spectrum for
periodicities less than 26 years was dominated by a peak at 22 years. This
behaviour tallies closely with the known behaviour of the 'double sunspot'
cycle (see Section 6.1). But an analysis of global temperature records, including land-based observations, by J. B. Elsner of Florida State University and
A. A. Tsonis of the University of Wisconsin, concludes that the bidecadal
oscillation is only present if the data before 1880 are included.[9] For the period 1891 to 1990 there is no significant evidence of this cycle. In contrast,
the most important periodicity is around 5 to 6 years, which is attributed
to El Niño Southern Oscillation (see Section 5.4).

[6] Labitzke & van Loon (1999), Chapter 4.   [7] Mock & Hibler (1976).
[8] Newell *et al*. (1989).   [9] Elsner & Tsonis (1991).

3.3 Rainfall records

As in the case of temperature records, the British Isles has some of the lengthiest and most comprehensive rainfall statistics in the world. But for a variety of reasons there has been much more extensive analysis of the rainfall records from around the world than in the case of temperature records. In part, this is because the economic and social impact of year-to-year rainfall fluctuations can be so much greater than temperature variations. So the search for cycles has had a greater sense of urgency. While the British records provide a good point to start from, the claims of cycles from different parts of the world will form a much greater part of the analysis of rainfall records.

The best known British rainfall series is a composite England and Wales series.[10] This series is the product of the work of the many amateur observers who kept records throughout the eighteenth and nineteenth centuries and, in particular, the subsequent work of one individual. The rigorous study of rainfall in England began with G. J. Symons, who published the meteorological journal that was the forerunner of the *Quarterly Journal of the Royal Meteorological Society*. Largely through his efforts in setting up the British Rainfall Organisation and the journal *British Rainfall*, the UK has one of the longest, most extensive and most reliable rainfall data sets. His assiduous work in collecting together earlier amateur observations laid the foundation of the England and Wales series. Subsequent work by a number of researchers led to a series which extends back to 1727 and which is now kept up to date by the UK Meteorological Office. In addition, more recent work has defined lengthy records for specific sites. These include one for Kew, which goes back to 1697, one for Pode Hole, Lincolnshire, from 1726, and one for Manchester from 1786. While there is considerable doubt about the long-term trends of these records in the eighteenth century, it is probable that fluctuations of shorter period will not be too badly affected. So cycles with wavelengths less than about 25 years can be regarded as relatively undisturbed.

A study by R. Tabony of the UK Meteorological Office of these series using both spectral analysis and filtering techniques provides yet more evidence of the complexity of cyclic behaviour.[11] The spectral analysis was conducted on detrended data for different parts of the year, including the

---

[10] Nicholas & Glasspoole (1932) and Wigley, Lough & Jones (1984). This monthly series is now routinely updated by the Hadley Centre and available on its website: http://www.met-office.gov.uk/research/hadleycentre/obsdata/index.html.

[11] Tabony (1979).

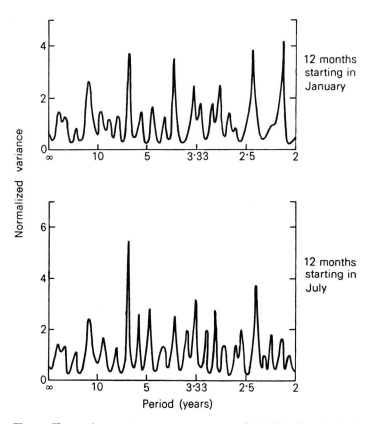

Fig. 3.2. The maximum entropy power spectrum of rainfall at Kew, England, summed over 12 months in January (top), and 12 months starting in July (bottom) during the epoch 1697 to 1975. (From Tabony, 1979. With permission of the Controller of Her Majesty's Stationery Office.)

winter and summer half-years and the conventional three-month seasons, together with the 12 months starting in January, April, July and October. Of the features common to all four series, the QBO periods of around 2.1 and 2.4 years stand out best. The latter is the most striking but is essentially a feature of the summer half of the year. Furthermore, the choice of starting point makes a difference. In the case of the Kew series, if the data are summed over the 12 months starting in January, which preserves the summer half-year, a large quasi-biennial peak is detected. But when the series is summed over the 12 months starting in July, which divides the summer half of the year, the 2.1-year cycle disappears (Fig. 3.2).

Other periodicities of note include a 3.9-year cycle in annual and winter rainfall, which is most evident in the record for England and Wales, but is also visible in the other series. Rainfall summed over the winter half-year is evident in the Kew, England and Wales, and Manchester series. This is

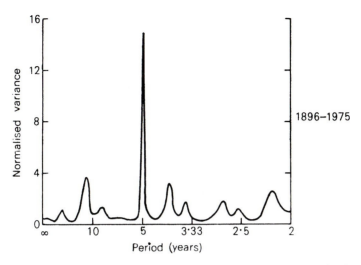

Fig. 3.3. The maximum entropy power spectrum of rainfall over England and Wales during the epoch 1896 to 1975, showing a pronounced periodicity at 5 years. (From Tabony, 1979. With permission of the Controller of Her Majesty's Stationery Office.)

also some evidence of a 50-year cycle in the England and Wales series. But, interestingly in terms of the overall survey of cycles, there is no significant evidence of the 11- and 22-year periodicities that might be associated with the sunspot cycles. The nearest approximation is a 12.6-year cycle in the annual and summer half-year Kew data.

When the series are split up into 80-year epochs to check whether the observed cycles persist throughout the entire record, some more confusing results emerge. The most important is that no spectral peak reached the 5% significance level in all the epochs examined. Indeed, there are large differences in the power spectra for the different epochs. This suggests that rainfall in Britain is dominated by random fluctuations. None the less, there are some interesting features. For instance, the 5-year cycle in the winter half-year spectrum for the period 1896 to 1975 (Fig. 3.3). Moreover, there is some evidence of a cycle of about this period in earlier epochs. This temporal variation can be seen even more clearly when the series for rainfall summed over 12 months starting in October is smoothed using a unitary filter (see Appendix A.6) centred on 5 years (Fig. 3.4). The fluctuation is well developed between 1860 and 1885 and after 1925. But at other times the fluctuation is either absent or much less evident.

The overall conclusion from this lengthy set of rainfall data is that, as with the central England temperatures (Section 3.1), the same enigmatic features emerge. Apparently significant periodicities are present only at certain times of the year. Moreover, they come and go throughout the record

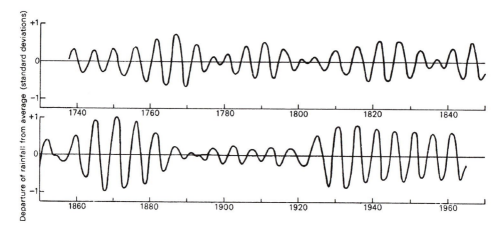

Fig. 3.4. The time series of England and Wales rainfall summed over 12 months starting in October, filtered with an 11th-order filter which illustrates how the 5-year periodicity comes and goes over time. (From Tabony, 1979. With permission of the Controller of Her Majesty's Stationery Office.)

in a way that is difficult to explain. But worse still they are not the same frequencies that are observed elsewhere. So, apart from adding to the body of evidence for the QBO, the picture is further complicated. Finally, these uncertainties are consistent with other work that reinforces the gloomy conclusion that none of the observed periodicities is sufficiently reliable to be used to produce meaningful predictions.

Despite these overall negative conclusions, it may not be correct to extrapolate them too far. It could be that the nature of rainfall over the British Isles is less sensitive to external perturbations than the much more variable seasonal and annual rainfall in other parts of the world. Even in Britain there is some evidence that extreme events may show a greater propensity to exhibit cyclic behaviour. An example of this is an analysis of annual figures of extreme one-hour rainfall in the UK in the period 1881 to 1986. This shows periodic variations of lengths of approximately 7, 11, 20 and 50 years.[12] These results may have some implications for assessing extreme rainfall, and hence the chances of damaging floods.

3.4 Chinese rainfall

Although the British rainfall series are the longest available records, based on direct instrumental measurements, there are a number of other sources of information about rainfall that go back somewhat further into the past.

[12] May & Hitch (1989).

Table 3.1. *Periodicities in Chinese droughts and floods*

Period (years)	Region of its predominance
80–160	The Yellow River region
36	The east part of the Yang-tze River, and the south-west of China
22	The north-east of China, and the central part of the Yang-tze River
11	The Yellow River region, and the south-east of China
5–6	The south of China
QBO	The Yang-tze River region, and north of the Yellow River

Note: QBO, quasi-biennial oscillation.

Of particular interest are the series that have been constructed drawing on China's huge store of historical records. Researchers at Beijing University have combed through these records to produce a thorough analysis of events going back to the fifteenth century.[13] So extensive are the references that in each year during the last five centuries they have graded droughts or floods for over 100 locations covering the whole of east China. These results lack the precision of instrumental observations, but because of their prosaic quality they contain considerable information about the prevailing weather. For example, in 1560 when both the north and the south of the country experienced drought, the Yang-tze basin had floods. The records include such items as in Dutong 'men were eating men', in Shijiazhung there was no rain in spring and summer and in Beijing there was a locust pest. In the south there was no rain in June and July in Jinhua and drought in late summer in Liuzhou. By way of contrast, there were floods in Nanking, and Yichang was overwhelmed with floods (see Table 3.1).

A group at the State University of New York has performed a spectral high-resolution fast Fourier transform analysis of the record for wet and dry years in Beijing for the period 1470 to 1974.[14] This concludes that the record is dominated by long-period fluctuations. The most significant feature is an 84-year cycle that is significant to the 0.1% level. There are also strong peaks at 126 and 56 years whose significant level is 1%, as is that of a peak at 18.6 years. There is also a marked peak at 9.9 years, but there the 11- and 22-year cycles are either weak or non-existent. These results have, however, been criticised by the researchers at the Climatic Research Unit of the University of East Anglia, who have extended the analysis to rainfall records within 600 kilometres of Beijing.[15] They conclude that although there are

[13] Wang Shao-Wu & Zhao-ci (1981).   [14] Hameed *et al.* (1983).
[15] Clegg & Wigley (1984).

a number of periodicities, they are generally unstable in space and time and so have little or no physical significance and have negligible predictive value.

More generally, a review of research to identify lower-resolution spectral analyses of records for different parts of the country for both the long-term historic records and the instrumental records since the late nineteenth century[16] provided broad confirmation of the negative results obtained by the University of East Anglia group. This showed a similar bewildering array of different periodicities from place to place, which also mirror the confusing picture in other parts of the world. The most important of these periodicities are set out in Table 3.1.

3.5 US rainfall and temperature patterns

Of all the areas of possible longer periodicities in the weather, the 20-year cycle in drought in the Mid-West United States (see Section 1.1) is probably the most extensively investigated. This is because of two factors. First, the 'Dust Bowl' years of the 1930s and subsequent droughts in the United States have had major economic consequences both for the United States and for the rest of the world, given the part played by American grain production in the world market. Secondly, the geographical extent and the comprehensive nature of the meteorological records, both instrumental and proxy data, make it possible to conduct a much more thorough examination for the case for cycles. In this chapter we will concentrate on the instrumental records and in particular the work of Robert Currie of the Institute of Atmospheric Sciences at the State University of New York.[17] The proxy data will be discussed in Chapter 4. This separation not only reflects the different types of data that have been used but also intriguing differences in the results that have been obtained from them.

Currie's work using MESA techniques has shown clear evidence of cycles in corn production in Iowa (Fig. 3.5). The two most important features occur at periods around 10–11 and 18–20 years. The first could possibly be linked with solar variations and the second may be attributable to 18.6-year lunar tidal effects. As the proxy data also show evidence of solar effects, we will concentrate on the lunar effects here. This effect is not found everywhere. It shows up in 894 of a total of 1219 records of annual total rainfall that have been examined. Using the monthly data from

---

[16] See Dombros & Gongbing (1988), pp. 195–201.

[17] Currie (1981) and Currie & O'Brien (1988).

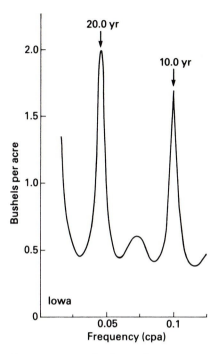

Fig. 3.5. An example of a maximum entropy spectrum of US crop production (corn in Iowa) showing clear evidence of periodicities at 10 and 20 years. (From Currie, 1987.)

these records reveals that the 19-year variation appears in 10 183 out of 14 628 of these records.

When the records are examined using statistical filters, an additional complication shows up in the way in which the rainfall patterns are linked to the lunar cycles. For example, the rainfall records from Pennsylvania to New England show that during the nineteenth century the least amount of rainfall coincided with a maximum of the lunar tidal force in 1843, 1861 and 1880. There is then a period of transition before a new pattern is established, with the maximum amount of rainfall coinciding with the peak lunar tidal force. The new pattern became established by 1917 and continued through successive maxima in 1936, 1954 and 1973. This switch in phase of 180 degrees of the periodicity is a well-known phenomenon in non-linear systems. But its unpredictability is a serious drawback in using evidence of cycles to underpin long-term forecasts. It is also a feature that can produce misleading results when conducting spectral analysis of time series. The filtered records also show that the 19-year wave varies in amplitude over time. It was roughly constant until 1940 and then increased significantly over the next cycle and a half to result in the major drought that gripped the north-east United States in the mid-1960s.

These complications are, however, small compared with the regional changes that occurred over 100 years within the study area. Whereas in 1880, the dry conditions covered almost all of the north-east states, for much of the first half of this century there were compensating areas of wet and dry. So while the average rainfall for the entire region showed clear evidence of a 19-year cycle, its intensity varied substantially from place to place. It was only in the 1950s that it regained a spatial coherence, which then led to the drought in the 1960s that was so much more intense than previous dry spells.

Currie also conducted extensive studies on US temperature patterns.[18] These studies also provided substantial evidence of two principal cycles with periods of around 10 to 11 years and 18.6 years, which he attributed respectively to solar and lunar influences. These effects are most striking east of the Rocky Mountains and north of latitude 35° N. More recent work by Peter Thejll, at the Danish Meteorological Institute, has explored 10- to 11-year ('decadal') cycles in global land air temperature records.[19] He partly confirmed Currie's conclusions, in that the most significant feature of the global pattern of the decadal signal is a marked cooling over North America, east of the Rocky Mountains, when sunspot numbers are high. Other features of this analysis is that there is an overall slight global warming with high sunspot numbers, with the most significant warming being off the west coast of North America, near the Black Sea, eastern Manchuria, and particularly in the mid-Atlantic off the west coast of Africa. In spite of the high significance of the pattern over North America, which, in its high-sunspot form, bears a marked similarity to the conditions that bring cold winters to eastern North America (see Section 3.10), Thejll is at pains to stress that this is not sufficient reason to conclude that the link between the 11-year solar cycle and global air temperatures is proven: statistically, it would not be that unexpected to find somewhere around the world that exhibited such a correlation. The important question is whether the decadal features originate from some external forcing, which may be capable of producing differing regional impacts, or are the product of the natural internal variability of the climate system.

3.6 Nile floods

Not all meteorological data are the product of standard instrumental records. For instance, the observation of water levels during the wet season

[18] Currie (1993).   [19] Thejll (2001).

can provide a direct measure of rainfall in the catchment area drained by a river. Of all the places where the flood level following the rainy season is measured, the most important must be in the Nile valley. Since prehistory the annual flood due to the summer rains in the highlands of Abyssinia and equatorial Africa has been the lifeblood of agriculture along the Nile. Moreover, given the introductory comments in Chapter 1 of the 7 years of plenty and the 7 years of dearth, it would be interesting to discover whether the behaviour of the Nile did indeed exhibit cyclic behaviour.

It is possible that records were kept of Nile flood levels well before the unification of Upper and Lower Egypt around 3150 BC. The earliest known records date, however, from the Early Dynastic Period. Carved on a stone stele during the Fifth Dynasty (*c.* 2400 BC), they include the height of the flood for every year back to the reign of King Djer around 3090 BC. But the most reliable set of measurements date from AD 622 onwards. These are continuous up to AD 1470 and then, apart from a few gaps, run up to modern times. As such they are the longest annual climatic time series monitoring the rainfall in a large drainage basin. They show major long-term fluctuations with periods of low discharge between AD 930 and 1071, and AD 1180 and 1350. High discharge episodes occurred from AD 1070 to 1180 and from AD 1350 to 1470.

As for cycles, a fast Fourier transform analysis of the record has shown that the most significant spectral feature is a peak with a period of 77 years.[20] This contains 3.5% of the variance and is significant at the 0.1% level. Two other important peaks are at 18.4 years, significant at the 0.5% level, and 53 years, significant at the 1% level. In addition, there are other interesting features in the range 17 to 47 years. Analysis of later data (AD 1690 to 1962) has confirmed that there is a significant feature at about 18.6 years that could be linked to lunar tidal effects (see Section 6.2). But, as far as the Genesis story is concerned, there seems to be no evidence of a 14-year cycle which could be linked with the 7 years of plenty and the 7 years of dearth. So, either what Joseph foresaw was an isolated event, or, if it was part of a more regular variation, the period changed between the end of the Thirteenth Dynasty (around 1700 BC) and the behaviour of the Nile since AD 622.

A more recent study, by hydrologists Elfatih Eltahir and Guiling Wang[21] of the Massachusetts Institute of Technology in Cambridge, Massachusetts, has correlated Nile river heights with records of Pacific Ocean temperatures dating back to 1872. This work suggested that about 30% of the Nile's annual fluctuations in water level could be linked to El Niño (see Section 5.4). This warming of the equatorial Pacific tends to cause droughts in parts of

[20] Hameed *et al.* (1983).　　[21] Eltahir & Wang (1999).

Africa with tributaries that feed the Nile, lowering its level. Cooler episodes (known as La Niña) tend to produce higher rainfall in these watersheds. These observations indicate that Nile records may provide some insight into El Niño, at least as far back as AD 622.

3.7   Pressure patterns

Many early searches for evidence of cycles in climatic records focused, however, on atmospheric pressure measurements or often on change in the difference in pressure between different parts of the world. This approach recognised the limitations in many meteorological data and was built on the widespread nature and inherent quality of surface pressure observations. Using averaged pressure readings avoided some of the extreme fluctuations that occur in other variables, notably in rainfall records. This reduced variability made it possible to detect 'real' cycles more easily. Furthermore, the examination of the differences in pressure between widely different locations has the advantage of concentrating on shifts in global weather patterns. So, by looking at the bigger picture, analysis of longer-term pressure statistics implicitly considered the slowly varying nature of certain climatic patterns and effectively included the role of the oceans in establishing longer-term atmospheric pressure patterns. Moreover, where subsequent attempts were made to link periodic variations in these variations with external influences (e.g. sunspots or astronomical perturbations), evidence of global patterns being disturbed by these physical effects tends to be more convincing than if the influence is merely localised.

The most intriguing feature about the best known of the pressure patterns is how values see-saw between different parts of the world. Often termed 'oscillations', these regular variations have a lengthy meteorological history. They have been identified in many parts of the world and at different times of the year. In considering instrumental records we will concentrate on just three of these oscillations, which were identified by Sir Gilbert Walker and his colleagues in the 1920s and 1930s.[22] When we move onto the wider question of how the atmosphere and the oceans interact in Chapter 5, the analysis of these oscillations will, however, widen. The first two patterns relate to winter pressure and temperature anomalies over the North Atlantic and North Pacific and are effectively measurements of the strength of the westerly circulation in mid-latitudes of the northern hemisphere. The third

[22] Definitions of these oscillations are given in Montgomery (1940).

is the behaviour of pressure patterns over the tropical Indian and Pacific Oceans, which may play a crucial part in the global weather machine, and which is discussed in the next section.

Since the eighteenth century it has been recognised that when the winters in Greenland are unusually warm the winters in northern Europe are exceptionally cold and vice versa. As the missionary Hans Egede Saabye observed in a diary kept in Greenland during the period 1770–78: 'In Greenland all winters are severe, yet they are not alike. The Danes have noted that when the winter in Denmark was severe, as we perceive it, the winter in Greenland in its manner was mild, and conversely.' Similarly, a paper published in 1811 provided a list of the extreme winters in Germany and Greenland between 1709 and 1800 that clearly confirmed this seesaw effect.[23]

Early work showed that there was a tendency for pressure to be abnormally low near Iceland in winter when it is high near the Azores and south-west Europe. This pattern was defined as the North Atlantic Oscillation (NAO) and is usually measured in terms of the normalised pressure difference between the area of low pressure that usually exists over Iceland and the high pressure region in the vicinity of the Azores. It is of particular interest in the winter part of the year (broadly speaking from November to March), and is a measure of the westerly circulation over the North Atlantic. When the Iceland low is particularly deep and the Azores high is well defined then the circulation is strong and the NAO index is defined as 'positive'. Conversely, when the gradient is reduced or even reversed, with high pressure near to Iceland and low pressure near the Azores, the index is 'negative'.

When the NAO is in the positive phase it produces strong westerly winds over the North Atlantic, and the reverse pattern with much weaker circulation. The strong westerly pattern pushes mild air across Europe and into Russia, while pulling cold air southwards over western Greenland. The strong westerly flow also tends to bring mild winters to much of North America. One significant climatic effect is the reduction of snow cover, not only during the winter, but also well into the spring. At the same time sea ice cover in the Kara and Barents Seas and the Arctic Ocean decreases, while the ice cover in Baffin Bay/Labrador Sea, Hudson Bay and the Gulf of St Lawrence increases. The reverse meandering pattern often features a blocking anticyclone over Iceland or Scandinavia, which pulls arctic air down into Europe, with mild air being funnelled up towards Greenland. This produces much more extensive continental snow cover, which reinforces

[23] van Loon & Rogers (1978).

the cold weather in Scandinavia and eastern Europe, and often means that it extends well into spring as long as the abnormal snow remains in place.

Research in the late 1970s at the National Center for Atmospheric Research in Boulder, Colorado, and the University of Colorado confirmed that these early observations were part of a correlated pattern of pressure variations across the northern hemisphere. This showed that the pressure anomalies are so distributed that the pressure in the region of the Icelandic low is negatively correlated with pressure over the North Pacific Ocean and over the area south of 50° N in the North Atlantic Ocean, the Mediterranean and the Middle East, but positively correlated with the pressure over the Rocky Mountains. Since 1840 the see-saw, as defined by temperatures in Scandinavia and Greenland, occurred in more than 40% of the winter months.[24]

Since mid nineteenth century, the NAO has fluctuated appreciably on timescales from several years to a few decades (see Fig. 3.6). It assumed a strong westerly form between 1900 and 1915, in the 1920s and, most notably, from 1988 to 1995. Conversely, it took on a sluggish meandering form in the 1940s and during the 1960s, bringing frequent severe winters to Europe but exceptionally mild weather in Greenland. So it appears to be a more persistent phenomenon than interannual fluctuations in snow and ice cover. Various efforts have been made to use proxy records to extend the NAO series back in time, notably with Greenland ice core data and tree ring data from many land sites around the North Atlantic. While these provide some interesting insights into climate behaviour in the region back to the beginning of the sixteenth century,[25] there is considerable doubt about whether these fully capture all the details of the NAO.[26] One of the reasons for these doubts is that these proxy records provide annual figures that lose some of the detail of the wintertime fluctuations, which are the most important feature of the NAO.

It was not until the 1990s that the NAO really came back into play.[27] This was the result of two developments. The first was the growing interest in oscillations stimulated by the 1982/83 El Niño. The second emerged from the period of exceptionally mild winters in the northern hemisphere during the late 1980s and early 1990s. An analysis suggested that a significant part of the global warming since the mid 1980s was associated with these very

---

[24]   van Loon & Rogers (1978).

[25] Luterbacher *et al.* (2002).   [26] Schmutz *et al.* (2000).

[27] A large number of papers on the NAO have appeared in recent years. A good place to find out more about both the history of the investigation of the NAO and our current state of knowledge is Marshall *et al.* (2001).

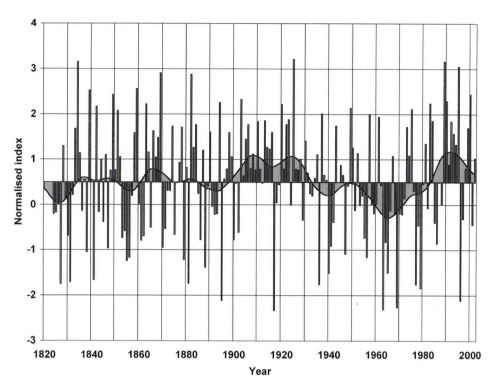

Fig. 3.6. The North Atlantic Oscillation in winter (December to February), based on the normalised pressure difference between Gibraltar and Reykjavik, provides a measure of the strength of westerly circulation in the mid latitudes of the northern hemisphere. (Data from the Climatic Research Unit, University of East Anglia.)

mild winters.[28] Indeed, since 1935 the NAO, on its own, can explain nearly a third of the variance in winter temperatures for the latitudes 20° to 90° N.

Spectral analysis of the various time series provides a complex picture. In the instrumental record since 1870 there is significant evidence of the QBO, especially in the first half of the record. For the rest, there is some evidence of a 7- to 9-year oscillation and a less-significant feature at around 14 years. But, overall only the QBO appeared regularly, and even this was not statistically significant at all station and grid-point pairs in the temperature and pressure data. Among the proxy records, spectral analysis also shows most of the variance as being concentrated in the frequency band with periods of less than 15 years.[29] Moreover, these oscillations are intermittent in nature, coming and going over the centuries. The weakness of these periodicities may, however, reflect the use of annual data that fails to capture fully the seasonal nature of the NAO, and suggests that it is characterised by

[28] Hurrell, J. W. (1995), (1996).   [29] Appenzeller, Stocker & Anklin (1998).

coherent phases when atmosphere–ocean interactions set up periodicities of between 5 and 15 years, and incoherent phases when the behaviour is random.

In addition to the shorter-term fluctuations in the NAO, there is considerable evidence of 65–80 year oscillation in temperature records for the North Atlantic and its bounding continents.[30] Although this basin-wide pattern of sea surface temperature anomalies is probably related to the NAO, it has now become known as the Atlantic Multidecadal Oscillation (AMO). This periodicity has also been identified in other records for the region, including sea level measurements, and has been linked with rainfall patterns over the United States.[31] It appears also to be part of a wider phenomenon that is seen as evidence of an oscillation in the global climate system of period 65 to 70 years.[32] The relative shortness of record (135 years) and the limited coverage in parts of the world, notably over the southern oceans and in the early part of the record, does mean, however, that this analysis should be treated with caution. Furthermore, this work finds mixed evidence for the bidecadal oscillation, with the strongest signal in the eastern equatorial Pacific and the South Pacific, while over South America, the Indian Ocean and the western equatorial Pacific either shorter interdecadal oscillations or the decadal oscillation prevail.

The other mid-latitude oscillation identified in the 1930s was the North Pacific Oscillation (NPO). This was a measure of the pressure difference between the high pressure are in the subtropical Pacific (usually measured in terms of surface pressure at Hawaii) and the low pressure in the vicinity of the Aleutian Islands (the Aleutian Low). This index attracted less attention at the time, and has only really come back into focus in the late 1990s (see Section 5.7). To complicate matters it has been reinvented in the form of the Pacific Decadal Oscillation (PDO) and linked with other interdecadal behaviour in the tropical and southern Pacific that have defined as the Interdecadal Pacific Oscillation (IPO).

3.8 The Southern Oscillation

'When pressure is high in the Pacific Ocean, it tends to be low in the Indian Ocean from Africa to Australia.' This is how Sir Gilbert Walker described in his papers in the 1920s and 1930s[33] what he named the Southern Oscillation

[30] Schlesinger & Ramankutty (1994).

[31] See for example Delworth & Mann (2000) and Enfield, Mestas-Nuñez & Trimble (2001).

[32] Schlesinger & Ramankutty (1994).

[33] Walker (1927), (1928), (1929) and Walker & Bliss (1933).

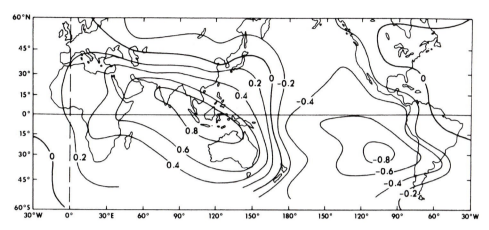

Fig. 3.7. The correlation of monthly mean surface pressure with that of Jakarta. The correlation is large and negative in the South Pacific and large and positive over India, Indonesia and Australia. This pattern defines the Southern Oscillation. (From Philander, 1983.)

(SO), and what has recently become perhaps the most intensely researched index of large-scale atmospheric pressure patterns. This is because of two factors. First, the SO is one of the most striking examples of interannual climate variability on a global scale. Over the tropical Pacific Ocean the SO is associated with considerable fluctuations in the rainfall, the sea surface temperature, and the intensity of the tradewinds, and it has been linked with extreme weather events around the world. The second, and related, factor is that the SO has become closely identified with El Niño events in the tropical Pacific (see Sections 5.4 and 7.3) – so much so, that the combined El Niño Southern Oscillation event has become widely known as ENSO in climatological circles.

Sir Gilbert Walker's original definition of the SO was based on the difference in pressure observations at Santiago, Honolulu and Manila, and those at Jakarta, Darwin and Cairo, together with figures for the temperature in Madras, rainfall in India and Chile, and the Nile flood. The subsequent increased information about pressure fields enabled the Dutch meteorologist Berlage to update the index in the 1950s.[34] He showed that the degree of organisation in the SO was truly impressive. Taking Jakarta as his reference station, he produced a map of the correlation of annual pressure anomalies (Fig. 3.7), which showed that the value at Easter Island had a surprisingly large value of −0.8. This map demonstrates that the SO is a barometric record of the exchange of atmospheric mass along the

[34] For an accessible version of this work and the subsequent analysis of these patterns see Bjerknes (1969).

complete circumference of the globe in tropical latitudes. Various studies looking for evidence of cyclic behaviour in the SO have used different indices to represent this phenomenon.[35] Sufficient to say, versions of these indices exist in monthly series that extend back to around the 1870s, and that the choice between these alternatives makes little difference to the analysis of cycles. In part, this insensitivity arises from the fact that the different versions of SO reflect efforts to normalise the index to encapsulate seasonal changes throughout the year. Because of the strong link between the SO and sea surface temperature anomalies there is, however, a marked persistence between the values observed in successive seasons (i.e. the climate has a 'memory'; see Section 2.7). This means that the analysis of the time series of the winter and summer halves of each year or of the annual figures produces almost identical results as does working with monthly values.

The standard view about periodicities in the SO is that while it has an average period of around four years, it is too irregular in nature, having intervals between major events ranging from 2 to 10 years, to identify cycles. Various attempts to identify more regular behaviour in the SO proved inconclusive, and the reason is apparent from recent work using wavelet analysis.[36] The power spectrum of the SO Index (SOI) has a broad peak between 3 and 8 years, and this peak has shifted around appreciably since the end of the nineteenth century (Fig. 3.8).

Earlier work at the CSIRO, Canberra, Australia, and the New Zealand Meteorological Service suggested a series of peaks at 3, 3.75, around 6, around 9, and 10 to 12 years, and sought to link these to fluctuations in rainfall around the southern hemisphere.[37] Rainfall in South America shows marked peaks at 3.75, 7 and 20 years; statistics for South Africa show weak fluctuations of periods of 16 to more than 20 years, of 10 to 12 years and of about 6 to 7 years. In the New Zealand figures there is a marked 10-year cycle, while the 20 and 6- to 7-year peaks are much less pronounced. But the filter analysis does suggest that there is some link between the observed cycles. The 6- to 7-year rainfall cycle in South America is out of phase with a similar cycle that appears elsewhere in the southern hemisphere from South Africa to Australasia. With the quasi-10-year cycle the situation

---

[35] The Southern Oscillation Index (SOI) is now defined in terms of the difference in pressure between Tahiti and Darwin, in Northern Australia. El Niño warm events occur when the index is negative and La Niña cold events when it is positive. The monthly values of the SOI published by different agencies (e.g. by the US Climate Prediction Center and the Australian Bureau of Meteorology) vary slightly, as they use different methods to standardise their data.

[36] Allan (2000).   [37] Vines & Tomlinson (1985).

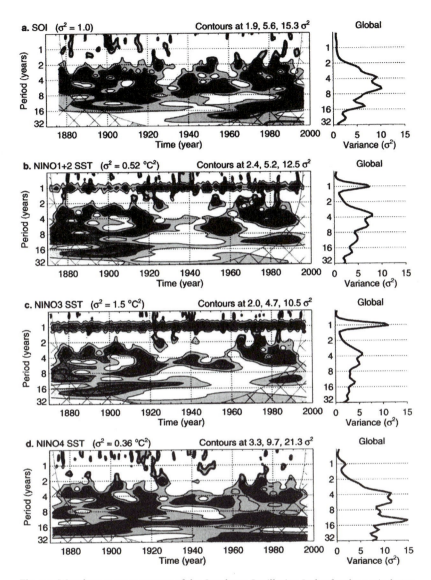

Fig. 3.8. Wavelet power spectrum of the Southern Oscillation Index for the period 1876 to 1996 using a Morlet wavelet. The contour levels are in units of variance and are chosen so that 50%, 25% and 5% of the wavelet power is above each level respectively. The thick contour shows the 10% significance level. The cross-hatched area indicates the 'cone of influence', where the variance is reduced by extending the series beyond the observed range using zeros as 'padding'. (From Allan (2000), Fig. 1.4a. in Diaz & Markgraf, 2000.)

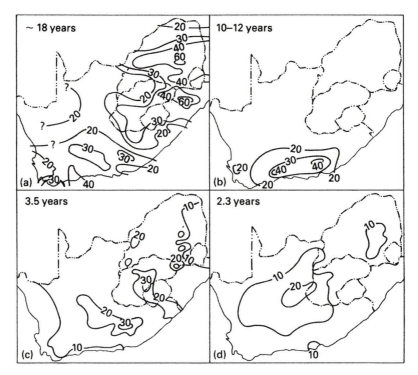

Fig. 3.9. The percentage variance in South African rainfall records over the interval 1910 to 1972 for significant oscillations with periods (a) around 18 years, (b) 10–12 years, (c) 3.5 years, and (d) 2.3 years, showing that the periodicity around 18 years is by far the most important, but that all the oscillations show significant geographical variation. (From Tyson, 1986.)

is more complicated: southern South Africa, south-eastern Australia and South America are in phase, whereas the corresponding fluctuations in north-eastern South Africa, Tasmania and New Zealand are exactly out of phase, although in phase among themselves. The 20-year cycle is essentially in phase in all areas.

More detailed studies of South African rainfall provide a slightly different and more complicated picture.[38] The most important periodicity in rainfall patterns during the period 1910 to 1972 is around 18 to 20 years, mostly around 18 years. Other important features are at 10 to 12 years, and around 3.5 years and 2.3 years. But all these periodicities show considerable geographical variation (Fig. 3.9). The quasi-18-year cycle is ubiquitous and in places highly significant. The 10- to 12-year cycle is seldom highly significant and only represents a significant part of the variance in Cape Province.

[38] Tyson (1986).

The 3.5-year cycle is ubiquitous and in places can contain a significant part of the variance, whereas the QBO is not particularly impressive.

3.9 Stratospheric winds

Having explained a wide range of surface weather records, many of which exhibit a quasi-biennial oscillation (QBO) plus a variety of other apparently periodic variations, we will now turn to what appears to be the best-established periodicity.[39] This is the regular reversal of the winds in the stratosphere over the equator. Westerly and easterly winds alternate in an oscillation of around 27 months that swamps completely all seasonal and lesser variations. This behaviour has been studied since the early 1950s. The period has varied from over 3 years to well under 2 years. It shows a series of well-defined characteristics. The wind regime propagates downwards as time progresses. The amplitude of the oscillation is greatest at an altitude of 30 km (pressure = 20 hPa). The easterly winds are stronger than the westerlies. The oscillation does not have a simple waveform (Figs. 3.10 and 3.11); instead, the time between the peak easterly winds and the peak westerly winds is less than the other way round. The velocity decreases as the height decreases. At high levels the easterlies last longer, while at lower levels the westerlies prevail longer. The amplitude between the extreme winds is some 40 to 50 m per second.

This regular behaviour is far better defined than any of the other fluctuations discussed in this chapter. But there is no adequate physical explanation as to how the QBO in the stratosphere is linked with all the fluctuations of similar duration that have been described earlier. Possible theories will be discussed in Chapters 6 and 7, but for the moment, in reviewing the evidence of the QBO and how it might be linked with some of the many cycles that have been identified so far, it must be remembered that we may be looking at two entirely different phenomena. This is because, in spite of the ubiquity of the QBO in surface weather records, the basic objection to perturbations propagating down from high in the stratosphere is that they have insufficient energy to modify the conditions in the turbulent denser lower atmosphere. In effect, the stratospheric tail is required to wag the tropospheric dog. For this to happen there would need to be some powerful non-linear amplification reinforcing the faint signals from above to enable them to influence our weather.

---

[39] A most thorough review of current knowledge of the QBO appears in Baldwin *et al.* (2001).

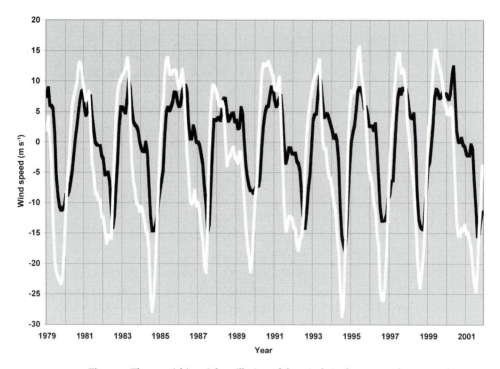

Fig. 3.10. The quasi-biennial oscillation of the winds in the stratosphere over the equator shows a pronounced periodic reversal. The scale of this oscillation is greatest at around 30 hPa (white line) and reduces at lower levels, as shown by the values at 50 hPa (black line), as the periodic feature migrates downwards over time, so that the peaks occur first at high levels. (Data from the National Atmospheric and Oceanic Administration (NOAA), available on the websites ftp://ftp.ncep.noaa.gov/pub/cpc/wd52dg/data/indices/qbo.u30.index ftp://ftp.ncep.noaa.gov/pub/cpc/wd52dg/data/indices/qbo.u50.index.)

3.10   Sunspots and the QBO

If the behaviour of the QBO in the stratosphere and its link with fluctuations of a similar period at lower levels is hard to explain, then the link between the stratospheric oscillation and sunspots is truly puzzling. Moreover, the way in which this link was identified is a good example of the problems that face investigators in unscrambling the complex behaviour of the global climate. As explained in Section 3.9, the QBO in the equatorial stratosphere reverberates through the stratosphere and down into the troposphere. But in so doing it is modified and distorted in a variety of ways.

It had been known since 1980 that the north polar stratospheric vortex during winter tended to be deeper and colder during the west phase of the QBO than in the east phase. But Karin Labitzke of the Free University of Berlin pointed out that at the solar maximum the polar stratosphere was

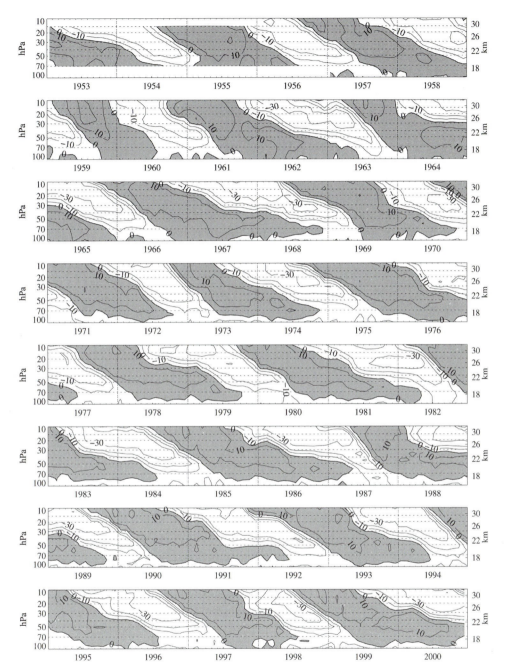

Fig. 3.11. A vertical cross-section of the stratospheric zonal wind from 1953 to 2000. Westerly winds are shaded. The zone of westerly winds moves steadily to lower altitudes and is replaced by easterly winds that in turn move down through the stratosphere. (With kind permission of K. Labitzke, Meteorological Institute, Free University, Berlin, Germany.)

unusually warm if the QBO was in its west phase. At an altitude of 22 kilo-metres (pressure 30 mb), warm winters with temperatures of about −54 °C, which feature a weaker more disturbed vortex, occur in west phases only when the Sun is at its most active. When the Sun is less active, winters as cold as −78 °C occur. In the east phase the opposite happens. Subse-quent work with Harry van Loon of the National Center for Atmospheric Research in Boulder, Colorado, showed how sorting out east-phase from west-phase winters transformed what looked like a complete muddle.[40] In short, the opposing effects of the solar cycle in opposite phases wiped out any correlation. When the two phases were considered separately, a dra-matic correlation appeared (Fig. 3.12). In the case of the west-phase years, there is a marked positive correlation with warmer winter periods when the Sun is active and colder winters when the Sun is least active. In terms of the polar stratospheric winter vortex this can be summarised as: when solar activity is high and the QBO is westerly the vortex is weak and disturbed, whereas easterly phases are accompanied by deep and undisturbed vor-tices. Conversely, when solar activity is low the vortex tends to be weak and disturbed when the QBO is easterly, but deep and undisturbed when it is westerly.

Standard statistical tests showed that there was a chance of less than 1 in 100 that the pattern in the west phase could happen by accident. But when combined with the pattern in the east phase, the chance of them both happening by accident was at most 4 in 1000. But it had only been observed for just over 3 decades and so there was still the possibility that it could be a coincidence. Indeed, some meteorologists were not convinced. To gain universal acceptance, it would have to continue to perform in the future (there is no prospect of it being extended back into the past, as there are no reliable records of stratospheric winds before the early 1950s). Secondly, an adequate physical explanation of the cause would have to be found.

At the time, the real interest centred on whether this combination of QBO and solar effects can be used to make useful predictions about the weather at lower levels. Labitzke and van Loon had examined the corre-lation between surface pressure anomalies and solar activity in the west-phase years for 19 winters. The results (Fig. 3.13) show that over northern Canada the positive correlation is as high as 0.7. This means that about half the variability of the sea-level pressure on the 11-year timescale can be as-cribed to solar influences. At a point in the western Atlantic (25° N, 55° W) there is an equal and opposite negative correlation. There is only about a 1 in 1000 chance that this pattern could occur randomly, so it looks as if there

[40] Labitzke & van Loon (1990).

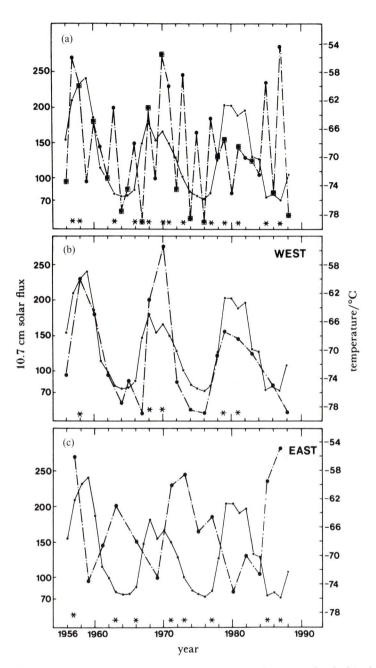

Fig. 3.12. The temperature of stratosphere at the 30-mb pressure level (altitude around 22 km) shows little correlation with solar activity – diagram (a). But when the data are separated into the two phases of the prevailing winds a striking pattern emerges – diagrams (b) and (c). (From Labitzke & van Loon 1990, with permission of the Royal Society.)

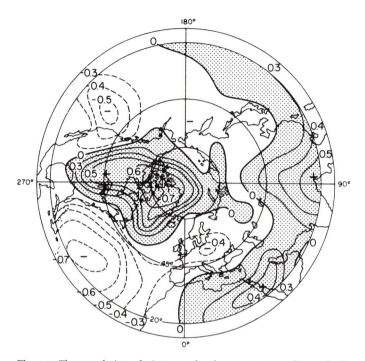

Fig. 3.13. The correlation of winter sea-level pressure anomalies and solar activity in years when the stratospheric winds are in the west phase. This shows that the pressure is abnormally high over northern Canada when solar activity is at a high level. The result is that cold northerly winds are more likely down the eastern coast of the United States while warm air is carried up into Alaska. (From Labitzke and van Loon 1990, with permission of the Royal Society.)

is strong evidence that solar activity is influencing winter surface pressure patterns in the northern hemisphere during west-phase years.

These observations led to the proposal of forecasting rules.[41] These were that when the Sun is at its most active and the stratosphere is in the west phase, the pressure will be higher than normal over North America, and lower than normal over the Pacific and Atlantic Oceans. Conversely, when the Sun is quiescent, the pressure will be abnormally low over North America and high over the adjacent oceans. Such anomalous pressure patterns (see Section 5.1) play a major role in extreme seasonal weather. When pressure is high over North America in winter, cold northerly winds will sweep down the eastern seaboard. But when the pressure is low over the continent and high over the oceans, warm southerly winds will flow up from the Gulf of Mexico. So in west-phase years we should expect to see cold winters on the east coast when the Sun is most active and mild winters

[41] Barnston & Livezey (1989).

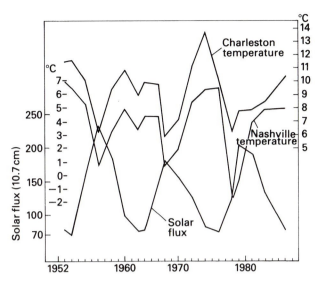

Fig. 3.14. A comparison of winter temperatures at sites in south-eastern United States (Charleston and Nashville) and solar activity in years when stratospheric winds were in the west phase. This shows that cold winters tend to occur when solar activity is high. (From Kerr, 1988.)

when it is at its quietest. As Fig. 3.14 shows, this is just what was observed at Nashville, Tennessee, and Charleston, South Carolina, between 1953 and 1986. These results may in part explain why winter temperatures in the eastern half of the United States at times show a marked biennial oscillation (see Fig. 1.3).

More generally, the change in pressure gradient between northern Canada and the mid-Atlantic during the solar cycle should affect the behaviour of depressions moving up the east coast and out across the North Atlantic. Again, the evidence of the west-phase years supported this expectation. The number of low pressure systems that crossed longitude 60° W between the latitudes of 40° N and 50° N in the west-phase years suggested periodic behaviour. The number was fewer than normal in years when the Sun was most active and higher than normal in quiescent years. In addition, latitude variation of winter (December to February) storm tracks in the North Atlantic showed significant variations. An analysis of storms north of 50° N between 1921 and 1976 showed that the average track was some 2.5° further south at sunspot maxima than at sunspot minima.[42] More striking was the fact that between 1952 and 1976 for west-phase years the peak-to-peak amplitude of the solar cycle variation increased to about 6 degrees of latitude.

[42] Tinsley (1988).

These observations appeared to amount to a significant climatic effect. Not only did they provide a potential forecasting tool, but they also offered insight into other phenomena, such as the 'see-saw' effects in winter temperatures between Greenland and Europe discussed in Section 3.7. They might also hold the key to other observations which point to a potentially significant connection between various weather events in the Atlantic and solar activity. For example, there is evidence of the number of tropical cyclones ('hurricanes') and the length of the tropical cyclone season being related to solar activity.[43] Analysis of the data from 1871 to 1973 showed distinct evidence of an 11-year cycle in both variables, and some sign of a 22-year cycle, plus less impressive evidence of features at 9, 15 and 52 years.

But the QBO–sunspot theory ran into serious trouble in 1989 when it failed its first major forecasting test. Rapidly rising solar activity and a west-phase QBO pointed clearly to a severe winter in the United States from December 1988 to February 1989. Although February was cold, overall the winter was mild. In searching for an explanation of what went wrong, Anthony Barnston and Robert Livezey at the US National Weather Service's Climatic Analysis Center came up with an interesting explanation. It was argued that the expected pattern was thrown out of gear by events in the tropical Pacific. When the sea surface temperatures are abnormally high – El Niño (see Section 5.4) – the chances of a cold winter over North America are increased. Conversely, well below normal temperatures in the equatorial Pacific – La Niña – should produce the reverse effect. In 1989, for the first time since the early 1950s, a La Niña coincided with high solar activity, and a westerly phase QBO.

Barnston and Livezey proposed that La Niña effectively cancelled the effect of the QBO plus high solar activity. At the end of 1990, everything appeared to be coming together. The Pacific appeared to be warming in the autumn. But the computer models of El Niño (see Section 5.5) told a different story. In the event, the models performed better. The progression of the winter did not go according to plan. December 1990 was exceptionally cold across much of the western United States, but in the east it was above average. The January 1991 pattern was less extreme, although the eastern half was slightly above average while much of the western half of the country was a little colder than normal. As for February, over the country as a whole it was the third warmest February of the century. Overall, this result was most disappointing. When combined with the failure in 1988/89 and a successful winter forecast for 1991/92 based on computer models of El Niño it suggests that at best the QBO–sunspot effects are a minor factor to winter weather

[43] Cohen & Sweetser (1975).

patterns and that events in the tropical Pacific, plus the natural variability of the global weather system, are more important. At worst they would have to be consigned to the scrap-heap – yet another example of a potentially useful cycle that, having been identified and used to produce a forecast, promptly disappears.

Over the last 10 years it is probably true to say that in terms of seasonal forecasting the QBO–sunspots theory has been in limbo. In its place the ENSO connection has become the holy grail of seasonal forecasting (see Section 5.6). But, as will be seen in due course, progress in this area has been elusive with what look like highly successful forecasts being followed by far less impressive results. As for the connection between North American winters and the QBO plus solar activity, the events during the last sunspot-number peak were possibly the last nail in the coffin of this specific relationship. After a very mild winter in 1997/98, which was widely forecast on the back of the record-breaking ENSO warm event, there were two more even milder winters in the United States. These winters coincided with La Niña conditions in the equatorial Pacific. But, to rub salt into the wound, the second (1999/2000) was the warmest in a record going back to 1895, although it followed a year when the QBO had firmly in the westerly phase and Sun was approaching a peak in its activity. Then in late 2000, with the QBO moving into an easterly phase, the eastern United States experienced the coldest November and December since 1895. Finally, in 2002, although the QBO had not really swung back into the westerly phase, but solar activity was still high, came another exceptionally mild winter. Although the combination of lingering La Niña effects in the Pacific and global warming could be invoked to explain some of this behaviour, there was little sign here of the QBO–solar signal in tropospheric weather patterns.

As for the more general question of stratospheric behaviour, researchers continue to chisel away at the knotty problem of links between the stratosphere and both solar activity and tropospheric weather.[44] The most convincing signals have continued to emerge from the work of Karin Labitzke and Harry van Loon.[45] Drawing on the NCEP/NCAR global reanalysis of meteorological data, they have produced results covering the last four solar cycles. They conclude that there is a solar signal in both hemispheres. The signal is strongest in northern summer (Fig. 3.15). Also shown is the timing of the major volcanic eruptions of Agung in 1963, El Chichon, in 1982 and Pinatubo in 1991, all of which had a separate and significant warming effect on the stratosphere. In the northern winter the unstable dynamics of the cyclonic vortex in the northern hemisphere counteracts the solar influence

---

[44] Baldwin *et al.* (2001).   [45] van Loon & Labitzke (2000) and Labitzke (2001).

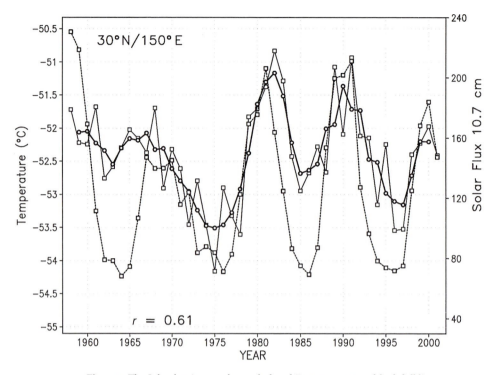

Fig. 3.15. The July plus August detrended 30-hPa temperature (black full line, squares) and the 3-year running mean (grey full line, circles) at grid point 30° N, 150° E, and the 10.7 cm solar flux, which is a proxy for ultraviolet variability (dashed line, squares). The major volcanic eruptions of Agung, El Chichon and Pinatubo, which occurred in 1963, 1982 and 1991, caused a substantial warming of the stratosphere. (With kind permission of K. Labitzke, Meteorological Institute, Free University, Berlin, Germany.)

in the west years of the QBO. The highest correlations between the solar cycle and stratospheric heights move poleward from winter to summer in both hemispheres, while the highest correlations with stratospheric temperatures move from one summer hemisphere to the other. These results are consistent with the modelling work (see Section 6.3) that indicates that the high correlations with the solar cycle are an indirect, dynamic result of changing levels of ultraviolet radiation from the Sun and the consequent changes in ozone concentrations in the lower stratosphere.

Attempts to link stratospheric behaviour with tropospheric weather patterns have focused on a wide variety of connections between the annual cycle, the QBO, ENSO and solar activity. In particular, the possibilities of 'phase-locking' between the annual cycle and the QBO and ENSO signals (see Sections 5.4 and 5.6), plus the potential photochemical consequences of solar activity on the stratosphere (see Section 6.3) have attracted considerable attention. This work may not offer the prospect of the major advances in long-range forecasting that appeared so enticing in the 1980s. Recent work does, however, confirm that surface weather patterns are influenced by long-lived circulation anomalies in the winter stratospheric vortex over the northern hemisphere,[46] which are linked to the phase of the QBO and ENSO. This connection may lead to better monthly forecasts. For the rest, the value of this work is the insight it provides into the physical links that underlie seasonal weather patterns and longer-term climatic variations.

3.11 Shorter-term cycles

Although the focus of attention in this book is mainly periodicities greater than a year, there are certain aspects of shorter cycles that must be addressed. Considerable work has been done on the power spectrum of atmospheric turbulence and wind speeds. This shows two pronounced peaks around one minute and four to five days with a broad minimum around one hour. The first of these is the product of microscale turbulence, and falls well outside the compass of this book. The second reflects the movements of synoptic scale features (i.e. highs and lows). Although these weather systems are an integral part of the longer-term fluctuations, their characteristic frequency is of no direct interest here, although in the form of synoptic noise they will be touched on in terms of the role noise might play in stimulating cycles (see Section 7.7). In addition, the diurnal temperature cycle and the diurnal and semi-diurnal atmospheric pressure cycles do not need to be considered. Possibly of more interest is a semi-diurnal variation in rainfall that appears to be linked to the lunar–solar gravitational tides.

On a slightly longer timescale, relatively little work has been done on periodicities from a week or two to several months. The reasons are complicated. In part, it is because the combination of synoptic fluctuations in the weather and the annual march of the seasons make these shorter-term cycles more difficult to detect. A second problem is that the higher harmonics of the annual cycle also mask other periodicities. In particular, the six-month (semi-annual) cycle is a common feature of certain meteorological behaviour (e.g. the zonal transport of momentum associated with the annual movement of the Intertropical Convergence Zone (ITCZ) and the associated subtropical high pressure belts). This movement of the ITCZ turns up in an even more basic form where its passage back and forth across the equator can produce two pronounced peaks in the annual rainfall statistics

---

[46] Thompson, Baldwin & Wallace (2002) and Baldwin *et al.* (2003).

(e.g. in Sri Lanka). More important is the possible 'phase-locking' of ENSO events to the movement of the ITCZ (see Section 5.6).

When it comes to fluctuations shorter than a few months, because so many studies have concentrated on monthly statistics, it is inevitable that some such fluctuations may have been missed. Nonetheless, the significant evidence of a monthly lunar–solar tidal cycle in rainfall statistics implies that this subject deserves more attention, as there appears to be a real tidal influence on individual weather systems. So Theophrastus may yet be proven correct.

There is, however, one short-term cycle where interest has burgeoned of late. The advent of satellite measurements in the 1960s led to the discovery that waves of cloudiness develop every 40 to 50 days in the Indian Ocean.[47] These intensify and sweep eastward across the Pacific at 30 km per hour and peter out before reaching South America. These waves can set the rest of the global atmosphere pulsating, with consequences for weather patterns at higher latitudes. These observations tie in with studies in the early 1970s of winds in the equatorial stratosphere that also show a similar periodicity. This work led Roland Madden and Paul Julian at the National Center for Atmospheric Research (NCAR) at Boulder, Colorado, to propose a mechanism linked to wind and pressure variations in the troposphere. They suggested that an area of convective activity moving eastwards in the tropics would disturb the boundary between the stratosphere and the troposphere. This disturbance might then propagate around the globe to trigger off the next oscillation. But their proposal received little attention until the early 1980s, when the satellite data provided evidence of the waves of cloudiness in the troposphere. During the 1990s, it became clear that these oscillations were central to a number of features of the global climate, including the timing and strength of ENSO events. In recognition of their pioneering work, this phenomenon is now known as the Madden–Julian Oscillation (MJO), although its generic nature sometimes leads to it being referred to as 'intraseasonal' oscillations.

The MJO is important because it significantly affects the atmospheric circulation throughout the tropics and subtropics, and also strongly affects the wintertime jet stream and atmospheric circulation features over the North Pacific and western North America. As a result, it has an important impact on storminess and temperatures over North America. During the summer the MJO has a modulating effect on hurricane activity in both the Pacific and Atlantic basins. Thus, it is essential to monitor and predict MJO activity, since it has profound implications for weather and short-term climate variability through the year.

[47] Madden & Julian (1971) and (1972).

The waves in the stratosphere appear to play an integral part in the theoretical explanation of the QBO in the equatorial winds (see Section 7.3). In addition, there appear to be links between the intraseasonal behaviour of the tropical troposphere and weather at high latitudes. In some winters these tropical fluctuations are mirrored by variations in the path and strength of the jet stream over East Asia, the North Pacific and North America. The influence of the tropical oscillation is also detectable in the strength of the summer monsoon over India. When a band of cloudiness coincides with the build-up of the monsoon, it tends to reinforce the convection process. Conversely, an intermediate clear region can slow the onset or interrupt the progress of the monsoon. Although this effect is superimposed upon the much stronger underlying drive of the monsoon, which is linked to the heating up of the Tibetan Plateau during the summer, it does appear to have a significant impact on the overall strength of this annual weather event.

The wider links have been confirmed in a spectral analysis of the winds over Singapore at an altitude of around 15 km between 1960 and 1985.[48] This shows that as well as a strong peak around 50 days there is a pronounced annual cycle and a marked QBO peak. This reflects the general observation that the 50-day oscillation is stronger during the northern winter than in the summer. It also suggests that there is a link between the short-term tropospheric cycles and the stratospheric QBO. This supports the idea that the QBO in the upper atmosphere may act as a trigger for the equivalent periodicities at lower levels.

In terms of topical weather patterns, the MJO is characterized by an eastward progression of large regions of both enhanced and suppressed rainfall, observed mainly over the Indian Ocean and the Pacific Ocean. The anomalous rainfall usually becomes apparent over the western Indian Ocean, and remains evident as it propagates over the warm ocean waters of the western and central tropical Pacific. This pattern of tropical rainfall then generally becomes nondescript when it reaches the cooler ocean waters of the eastern Pacific but can reappear over the tropical Atlantic and the Indian Ocean. Each cycle lasts approximately 30–60 days.

3.12 Summary

Having reviewed some of the vast array of studies that have been conducted on instrumental records, it is now time to take stock. So far we have let the analyses speak for themselves and have not questioned whether any of the

[48] Ichi-Kuma (1990).

Table 3.2. *A summary of the most significant periodicities in meteorological records (other than the QBO)*

Source	Period (years)
Central England temperature	3.1 5.2 14.5 23 76
US east coast temperature	4.5 9 20
Global air temperature	22 65–70
Beijing rainfall	9.9 18.6 56 84 126
US rainfall	11 18.6
Nile floods	18.4 53 77
North Atlantic pressure	7–9 14 50–90
Atlantic hurricanes	9.3 11 15 22
Southern Oscillation	3 3.8 6 10–12 52
South American rainfall	3.8 7 20
South African rainfall	3.5 10–12 18

results are less significant than claimed. While such an uncritical approach could lead to the wrong conclusions when considering individual studies, here we will rely on the weight of a large number of investigations to make or break the case for many of the claimed cycles (see Table 3.2).

The results of the various studies cited in this chapter are summarised in Table 2. Plainly, almost every possible periodicity between 2 and 200 years has been observed with some degree of certainty in some meteorological records somewhere at some time or another. But relatively few appear with high frequency. Those that do include:

(a) *The QBO*   This is the most widely observed feature in both troposphere and the stratosphere, and must clearly be regarded as a real feature in almost all those meteorological records where it has been observed. While there is no doubt about its reality in the stratosphere, in the troposphere it is less reliable, in that the term QBO is used to describe any periodicity in the range 2.2 to 2.8 years. This is reasonable, given the fluctuations that have been observed in the original QBO in the stratospheric winds. Nevertheless, it does limit the utility of the observations because what looks like a small span of periods is a big chunk of the frequency range 0.45 to 0.35 cpa. This means that the different components of what is broadly defined as the QBO can move in and out of phase in relatively few years. So the potential value of using the QBO to forecast tropospheric weather fluctuations may be limited.

(b) *3 to 4 years and 5 to 7 years*   A considerable number of power spectra cited in Table 3.2 have a potentially interesting feature in one or other, or both, of these ranges. But only rarely are they highly significant, and often they either do not appear or are a transient feature of the records. Nevertheless, they are worthy of mention and, as we will see in Chapter 5, they may be related to quasi-periodic fluctuations in the tropical Pacific. An alternative but less likely proposal (see Section 7.1) is that they are higher harmonics of the 11- and 22-year solar cycles.

(c) *11-year sunspot cycle*   This is the most popular and in many ways the most enigmatic of the 'cycles'. While there are a huge number of claimed observations of the 11-year cycle, there are a great many studies that show that it is absent in other records. Moreover, some of the best documented examples of the cycle have proved to be transitory, lasting only two or three cycles. Nevertheless, on the basis of the evidence cited here it is clear that something with a period around 11 years is a common feature in many records. The real issue, as will become apparent in Chapters 5 and 6, is whether this is merely a natural oscillation of the global atmosphere–ocean system or is driven by solar variability.

(d) *20-year cycle*   This is probably second only to the QBO as the most commonly identified periodicity in meteorological records. At this stage, we will not reach any conclusions about whether it is due to lunar tidal effects, or to the double sunspot (Hale) cycle, or perhaps as part of the natural variability of the global atmosphere–ocean system. Until we have looked at all the evidence, all we can say is that on the basis of instrumental records the case for many aspects of the weather being modulated by a 20-year cycle appears to be formidable.

(e) *80- to 90-year cycle*   This is a much less frequent feature in the spectra that have been produced – in part because many of the records that have been examined are barely long enough to provide clear evidence of such a lengthy periodicity. It is, however, included here for two reasons. First, in a number of lengthy records (e.g. central England temperature and Nile floods) it is a strikingly strong feature. Second, there is a comparable periodicity in the behaviour of sunspots, which means that there could be a direct physical cause for the observed variation.

(f) *200-year cycle*   There is some hint in the longest records that there may be a periodicity of roughly this length. This is noted in part because a similar periodicity (∼180 years) is detected in the sunspot series and lunar tides, and in part for continuity, as these longer cycles will be the subject of greater attention in the next chapter when we consider proxy data.

So what we can now say is that thus far we have found seven rather ill-defined periodicities, only some of which are found in a high proportion of the records. Furthermore, they tend to come and go in a tantalising way in many time series. This might be said to provide a less-than-convincing case for the existence of real cycles but for two features. First, there is a general impression of a hierarchy of periods between 2 and 20 years, with each successive one being roughly the sum of the previous two. This suggests that there may be non-linear effects at work serving to produce overtones and beats between any of the real features such as the QBO, and possibly solar cycles in the weather. The second and related feature is the clear evidence of a direct link between the QBO and the 11-year solar cycle. So there are some signs of other periodicities worthy of further investigation and interpretation. But before we can examine the physical arguments for the reality of the cycles so far identified, we must explore the wide range of other sources of information about past variations of the weather. These offer the prospect of both providing additional information about the cycles identified in this chapter and extending the timescale of the periodicities.

Chapter 4

Proxy data

The dust of antique time would lie unswept
And mountainous error be too highly heap'd
For truth to o'erpeer.
Shakespeare (*Coriolanus*)

The analysis of indirect information about the weather conditions of the past has been particularly useful in establishing the case for shifts in the climate. Often termed 'proxy data', these sources occur in many forms and can include almost any form of physical behaviour that reflects the influence of the weather. Ideally, proxy data should record in some permanent form the consequences of seasonal and annual changes of one particular aspect of the weather. In practice, many of even the best records are far more complicated. They often contain information about a variety of meteorological variables. Furthermore, they may be influenced by the weather over a number of years if there is some cumulative effect such as the build-up or decline of groundwater reserves. Finally, there may be problems of disturbance of the records by other external factors, which remove much of the fine detail in the original records, or introduce long-term fluctuations that cannot be easily corrected for or removed from the record.

In this chapter we will concentrate on those records where at least clear annual figures are available, such as tree rings, cores from corals and glaciers (including the ice caps of Greenland and Antarctica), and certain lake sediments. Even these records have some of the drawbacks noted above. But at least they will enable us to examine the evidence in these data for cycles of the same frequencies as have been identified in Chapter 3. This restriction does, however, limit the scope of the study and if followed rigorously would eliminate some of the most intriguing and convincing examples of cyclic behaviour. So, having explored the best records, which provide the clearest evidence of periodicities in the range 2 to 200 years, we will then expand the study to consider a wider variety of data; this provides insights into not only these periodicities but also extends our analysis to much longer climatic fluctuations. But, in practice, it will be seen that these have only limited impact in providing evidence of cycles with periods less than a millennium.

At first sight, this negative view may seem surprising, given the evidence of climate change on these longer timescales. For instance, there is no doubt that between the sixteenth and nineteenth centuries parts of the northern hemisphere experienced a markedly colder climate, especially in winter – often called the Little Ice Age. Prior to this there appear to have been periods of several centuries with warmer and colder climates. But, for a variety of reasons, until recently the proxy records lacked the precision to unravel these relatively small climatic shifts with any certainty. So although the evidence of climate change on timescales from centuries to millennia remains relatively limited compared with that concerning shorter- and longer-term variations, we are now beginning to get a clearer picture of periodicities in this time frame.

Where proxy records have made a much more substantial contribution relates to the longer timescale and in respect of much larger changes in the global climate. In particular, this will enable us to look at the evidence for the cyclic behaviour of the ice ages in the last million years or so and for linking them with the astronomical motions of the Earth. The reasons for doing this are twofold. First, they are an important part of any study of climate change. Second, the physical reasons for such cycles are much better established than those for shorter periodicities. So they will provide a benchmark against which to judge the other observations.

4.1 Dendroclimatology

The study of tree rings to obtain climatic information (dendroclimatology) has a long pedigree. Andrew E. Douglass started the classic work in 1904. Having worked at the Lowell Observatory in Arizona, he had become interested in looking for evidence that the sunspot cycle affected the weather. When he took up a post at the University of Arizona in 1906, he developed his examination of the cross-sections of ponderosa pines in the Flagstaff area to demonstrate that the same pattern of broad and narrow rings was to be found in all the trees and that the pattern in the outer rings tallied with local rainfall records. What is more, he appeared to produce convincing evidence that rainfall in the south-west United States was related to sunspot number – high solar activity coinciding with high rainfall and vice versa.[1] His other major discovery was that by identifying clear sequences of ring thicknesses and by using older timbers that overlapped living trees it was possible to build up a longer record of the sequences of tree rings going back several thousand years: a fundamental contribution to archaeology.

Here we will not consider Douglass' early climatic studies in detail, as this work has been overtaken by modern investigations. The importance of his pioneering work is twofold. First, as a result of his lifetime's work on tree rings, right up to his death at the age of 94 in 1962, the Laboratory of Tree Ring Research at the University of Arizona has become the leading centre of dendroclimatological studies in the world. Many other groups have done important work building up tree-ring series that can go back thousands of years in such places as Chile, Germany, Japan and Northern Ireland, but it is in Arizona that perhaps the most interesting work has been done. The second is that Douglass' work showed that the links between the weather each year and the thickness of the tree ring in that year was no simple matter. It depends not only on whether the growth of a tree in a particular site is dependent on a single variable (e.g. rainfall), but also how a tree responds to a sequence of good and bad seasons.

As a first approximation it is only where a tree is close to some form of climatic limit that it will show a clear correlation with the limiting meteorological variable. For instance, the growth of trees in subalpine regions or on the edge of the tundra will be most obviously affected by fluctuations in annual temperature. But those growing at lower levels in arid regions are more likely to be affected principally by variations in ground moisture. Trees growing in the central climatic zone will be less sensitive to weather fluctuations and will show a more complicated response to both temperature and precipitation changes.

The subtle link between tree growth and weather conditions means that the simple measurement of ring width alone cannot be used for the quantitative investigation of weather cycles. For instance, while the lengthy series obtained from examining oaks in Germany provided intriguing evidence of periodic variations from year to year (see Fig. 1.1), efforts to show a quantitative link with specific meteorological variables has been less successful. Similarly, oaks that grew in Northern Ireland show convincing evidence of dramatic short-term periods of climatic deterioration over a few years,[2] but cannot be used for the analysis of cycles. Not unexpectedly, normal growth of mature trees in temperate latitudes is dependent on the

[1] Douglass (1919).   [2] Baillie (1995).

complete range of weather elements. So while broad rings do appear to coincide with 'good' summers, the definition of what constitutes good conditions for oak trees is much more difficult to pin down. Consequently, in the first instance we need to concentrate on those historical records where direct links between either temperature or precipitation have been established and a more comprehensive picture has been built up. Once we have explored these examples we will extend the analysis to more sophisticated investigations. This involves using data with not only a variety of samples from a given site to iron out the local effects on individual trees, but also a good geographical spread of observations to gain a better measure of regional climatic effects. In this context, the history of work in the western United States provides a standard against which to judge other dendroclimatological studies.

The work of Charles Stockton and David Meko of the Laboratory of Tree Ring Research at the University of Arizona, together with Murray Mitchell of the National Atmospheric and Oceanic Administration (NOAA) Environmental Data and Information Service at Silver Spring, Maryland, in the late 1970s[3] provides an excellent example of the care needed in examining tree-ring records. They built up a comprehensive picture of the precipitation in the western two-thirds of the United States since AD 1600. From this it has been possible to construct various series of the severity and areal extent of drought between Canada and Mexico and from the West Coast to the Mississippi River using between 40 and 65 tree-ring sites. Spectral analysis of three different sets of series with four different levels of drought severity showed that the most prominent, and only reliable, feature was periodicity around 22 years (Fig. 4.1). As can be seen, the variance near 22 years exceeds the 95% confidence level and in one case reaches the 99.9% level – convincing by almost any standard.

Not satisfied with this apparently clear evidence of cyclic behaviour, the researchers then conducted an exhaustive filter analysis on the series. By using two adjacent narrow filters centred at 20.6 and 24.3 years, they successfully demonstrated that the coincidence between the observed cycle and the double sunspot cycle ('Hale' cycle) was not simply the product of the statistical analysis. By comparing the time series obtained by using these two filters it was possible to check that the phase of the observed cycles in drought severity (Fig. 4.2) did not shift appreciably. This analysis demonstrated that the phase of the cycle was closely linked to the phase of the Hale cycle thereby providing additional support for the existence of a physical link between the two phenomena. The statistical significance of

---

[3] Mitchell, Stockton & Meko (1979).

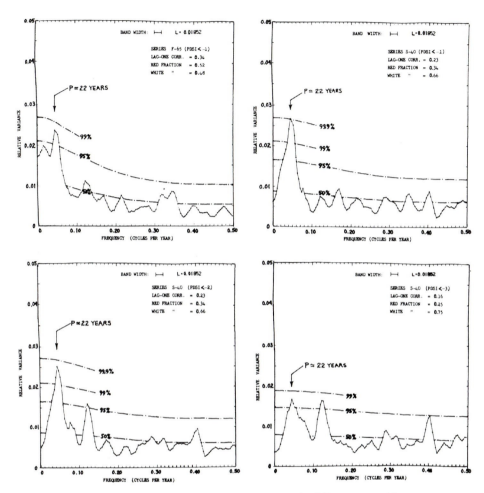

Fig. 4.1. Four examples of the variance spectra for different areas of the western United States for different levels of drought severity. These show that the dominant feature is a periodicity at around 22 years. The significance level is defined in terms of the assumption that the expected distribution of the variance would be 'pink' (see Section 2.7). (From Mitchell, Stockton & Meko, 1979.)

this link was assessed as being in the 5% to 1% range. This reinforces the suggestion that there is a strong association between the 22-year periodicity and the Hale cycle. In addition, there was some evidence that the episodes of drought exhibited a longer-term variation that appeared to match the 90-year (Gleissberg) periodicity in solar activity. The cautious conclusion of this work, published in 1979, was that while not wholly reliable, the risk of drought somewhere west of the Mississippi is appreciably higher in years immediately following the minimum of the Hale cycle.

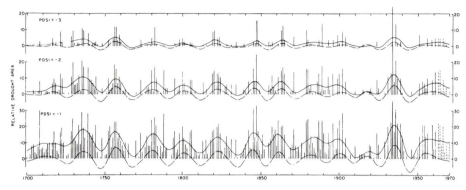

Fig. 4.2. Three examples of drought severity for 40 regions in the western United States for the period 1700 to 1970. The vertical bars denote the proportion of the areas experiencing drought at different levels from least severe (−1) to most severe (−3). The pair of wavy lines in each series shows the smoothed series using (in upper curves) a low-pass filter cutting off periodicities shorter than ∼12 years, and (in the lower curves) a narrow-band filter centred on 20.6 years (see Fig. 2.6). (From Mitchell *et al.*, 1979.)

An interesting feature of this work is that it appears to contradict the studies of US rainfall by Robert Currie (see Section 3.5), which concluded that the periodicity around 18 to 20 years was of lunar origin. Subsequent work on the drought index series, first by Murray Mitchell and then by Ed Cook, David Meko and Charles Stockton[4] has partially resolved this difference and at the same time provided additional insight into the complexities of the periodic behaviour in the weather. Murray Mitchell showed that although the 22-year cycle was the dominant feature in the series from 1600 to 1962, there was virtually no sign of the lunar cycle. But, if the series was cut in half and the two halves analysed separately then the lunar cycle appeared as strongly as the solar cycle. Indeed, between 1842 and 1962 it was markedly more pronounced than the solar cycle. The reason for this behaviour was that around 1780 the phase of the lunar cycle shifted by 180°, and so in the spectral analysis the signal in the first half of the series neatly cancelled out the signal in the second half. Moreover, the filter analysis did not show up this effect, as the filters used were not sufficiently sharp to discriminate between the two periodicities. Only with the detailed combination of spectral analysis and filtering was it possible to show that apparently there were both solar and lunar influences at work. As noted in Section 3.5, abrupt shifts in phase, as observed around 1800 in the 18.6-year cycle, are an indication of non-linear processes at work. This complex reaction to external periodic influences will be examined in Section 7.1.

[4] Mitchell (1990) and Cook, Meko & Stockton (1997).

The work by Ed Cook, David Meko and Charles Stockton confirmed there is a reasonably strong statistical association between the bidecadal maximum in the drought area rhythm and both the Hale solar cycle minimum and 18.6-yr lunar tidal maxima. The solar and lunar effects appear to be interacting to modulate the drought area rhythm, especially since 1800. They observe, however, that their results cannot distinguish between the possibility that the cycle in the drought area is the product of coupled atmosphere–ocean processes, or that it is the result of the solar and lunar forcing.

A great deal of work has been done on other tree-ring series. This is beginning to produce an increasingly interesting picture. Early work on annual growth rings of trees growing near the northern forest limit in Finland, where the dominant climatic control is likely to be summer temperature, produced some evidence of solar influences.[5] Features around 23, 90 to 100 and 200 years were of particular interest.

The existence of cycles around 100 and 200 years is supported by studies of tree-ring data from bristlecone pines (Fig. 4.3) at an altitude of 3384 m in the Campito Mountains of California.[6] These covered the period 3405 BC to AD 1885 and showed a marked similarity to the power spectrum for tree-ring widths and the variations in the concentration of the isotope carbon-14 (see Section 6.1). Of particular interest were the periodicities at 208 and 114 years. But, as so often is the case, the 208-year feature appeared to come and go throughout the record. There is also evidence of longer periods around 700 and 1400 years. Many other early analyses of tree-ring records from around the world produce a bewildering array of periodicities, many of which are neither highly significant nor reproduced in other studies. So, it is perhaps better to move on to more recent work.

More detailed information can be obtained from analysis of the structure of the wood within individual rings. Such parameters as the density of wood formed towards the end of the growing season (*maximum latewood density*), the corresponding *minimum earlywood density* and width of early and latewood growth can provide insight into weather fluctuations within individual seasons. For instance, stunted early growth could indicate late cold springs while lack of late growth could reflect cool, short summers. In recent years, particular attention has been devoted to the maximum latewood density, which has been shown to be a good measure of the growing-season temperature in regions that have cool moist summers. A network

[5] Lamb (1972), p. 235.

[6] Neftel, Oescheger & Suess (1981). Subsequent comparable lengthy analyses, including the Belfast oak record, are reviewed in Suess & Linick (1990).

Fig. 4.3. Bristlecone pine (*Pinus aristata*), native to mountainous areas in the south-western United States. These trees grow at altitudes above 3000 m. Because of their extreme longevity and the fact that they live near the limit of their climatic tolerance, they are uniquely important in dendroclimatological studies. (From Sherratt, 1980. Photograph Copyright © BBC, London.)

of nearly 400 tree-ring density chronologies has been built up around the world and has been used to construct regional climatic records across the northern hemisphere and for parts of the southern hemisphere.

In terms of weather cycles, recent work[7] by Keith Briffa of the Climatic Research Unit, University of East Anglia, and Fritz Schweingruber of the Swiss Institute of Forest, Snow and Landscape Research, on trees in northern Fennoscandia covering the period 1501 to 1980 (Fig. 4.4) has produced some interesting results, with the most prominent features being at around 11, 31 to 34, and 86 to 100 years. The shortest and the longest of these could be associated with solar variability. The middle one is a persistent feature of many tree-ring series, which suggests that variance with a periodicity of

[7] Briffa *et al.* (1990) and (1995).

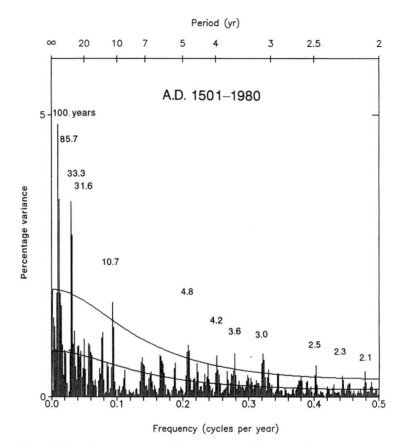

Fig. 4.4. The variance spectrum of the reconstructed Fennoscandian temperature series calculated from tree-ring data over the period 1580 to 1978, showing that the distribution is 'red' and that the most significant features are the periodicities at 85–100, 31–34 and 10.7 years. In addition, the various higher frequency features between 5 and 3 years, and between 2.5 and 2.1 years may be related to the El Niño Southern Oscillation (ENSO) and the quasi-biennial oscillation (QBO). (With kind permission of K. Briffa, Climatic Research Unit, University of East Anglia.)

between 30 and 40 years is reproducible feature of summer temperatures in high latitudes of the northern hemisphere.

One other aspect of tree-ring studies requires careful consideration. This is the fact that all analyses have to take account of the fact that as trees mature their ring widths naturally decrease, as does their maximum late-wood density. In the case of conifers this change tends to be exponential, but in broad-leaved species (e.g. oaks) it is a more complicated process. Standard practice is to calibrate growth curves for given species and then analyse the annual fluctuations about the normal. Where this process involves overlapping of series of tree samples, the age of which is not known, this standardisation process tends to iron out longer-term fluctuations and hence may produce misleading information about climatic variability. Often known as the 'segment length curse', this process can lead to tree-ring series having a suspiciously stable appearance over the longer term, and seriously underestimating the climatic variance on timescales of 50 years and longer and hence the significance of any periodic features.[8] The only way to avoid this pitfall is to restrict the analysis to wood of known age, but in the case of ancient timbers this rules out many samples which have no indication of age of the tree from which it was hewn and so greatly reduces the scope of climatic studies.

4.2   Varves

A varve is defined as a pair of thin layers of clay and silt of contrasting colour and texture (Fig. 4.5) that represents the deposit of a single year (summer and winter) in still water for some time in the past. The word comes from the Swedish for a 'layer' and reflects the fact that the original study of these sediments was the life's work of Baron Gerhard de Geer.[9] He examined the varves deposited in lakes formed by the retreat of the ice sheet that covered Scandinavia at the end of the last ice age. The term has now come to cover any collection of annual layers found in lakes, which can result either from changes in the rate of glacial run-off, or, more generally, changing climatic conditions that affect both the nature of sediments and the rate at which they form in lacustrine settings. As such they are a powerful tool for examining long-term climatic shifts and have the potential to provide independent evidence of weather cycles not only in recent centuries but also over a much longer geological span.

The scale of de Geer's work is worth recording. Starting with his fieldwork in 1878 he built up a picture of the regularity of glacial varves throughout Sweden. This monumental work continued until 1938, during which time many thousands of varves were measured for a large number of sites. It also needed careful detective work to match up these separate sets of observations, as in many cases the glacial formation of the varves had long

---

[8] Cook *et al.* (1995). More recent work by Cook and co-workers (Esper, Cook & Schweingruber (2002)) has shown that with careful calibration of lengthy tree-ring data sets from a wide variety of sites from around the northern hemisphere, it is possible to get a better measure of the longer-term climatic variability on timescales from decades to centuries.

[9] De Geer (1929).

Fig. 4.5. Varved clay deposited 10 000 years ago in Leppa Koski, Finland. Formed by the different rates of settling of sediment throughout the warmer months of the year when glacial meltwater enters lakes, these distinctive annual layers can be used to explore past climates. (From Selley, 1988.)

since ceased. This involved forming a set of overlapping series from different lakes in the same way as tree-ring series were constructed, so that the most recent varves in northern Sweden could be matched with older records from more southerly lakes. By recognising unambiguous sequences in different records it was possible to produce a year-by-year record of the retreat of the Scandinavian glaciers over more than 10 000 years. In principle, variations in these records should provide a measure of the weather, as thick varves should be a sign of warm weather and rapid melting. But, in practice, they are disappointing. Although they contain a great deal of information about major climatic shifts that occurred following the last ice age, the evidence of cycles is far less convincing.

Comparable work has been carried out on various lakes in the European part of the former USSR and in other parts of the world. These observations cover a wide range of sites and so variations in varve thickness could reflect changes in temperature, rainfall, evaporation and storminess. As a general observation, these records tend to have their most prominent periodicities in the ranges 2 to 3, 5 to 6, and 10 to 12 years. Perhaps the most striking example comes from a 4000-year record laid down at the bottom of two lakes in the Crimea that shows cyclic variations in thickness with a period length of 11.2 years and varying in individual cycles between extremes of 7 and 17 years.[10] These figures closely parallel sunspot behaviour (see Section 6.1). But overall, the results of analysing varves laid down over the last 10 000 years provide scant support for weather cycles. As with rings from trees growing in temperate latitudes, this failure may be the result of the varves reflecting the contradictory effects of a number of meteorological variables. For instance, fluctuations in temperature may tend to cancel out fluctuations in rainfall.

The importance of these results is that they enable us to interpret the significance of similar observations that have been made of various geological formations. There are plenty of examples of attempts to identify short-term periodic behaviour in geological varve records (evidence of longer-term fluctuations will be examined in Section 4.6). They cover a wide range of geological history from the Miocene to the Precambrian.[11] Frequently, they suggest the presence of the now-familiar periodicities (2 to 3 years, 5 to 6 years, and around 11 and 22 years). Less often, 90-year and around 200- and 400-year periods have been detected. But as a general rule none of these periodicities is particularly convincing, with the exception of the 11-year Precambrian cycle that appeared to be linked with solar activity (see Section 4.5).

---

[10] Lamb (1972), p. 252.   [11] Anderson (1961).

This pessimistic conclusion is supported by a thorough examination of the case for ancient weather cycles in the Eocene Green River Formation in Wyoming, Utah and Colorado. This formation was deposited 45 to 50 million years ago and consists of laminated organic-rich marlstones or 'oil shale'. It features alternate light and dark layers – light in early summer and dark in late summer/winter. The laminations are easily identified and statistical studies were conducted on three time series of 1469, 1869 and 4158 annual thickness measurements.[12] Fourier transform and MESA studies showed that the spectrum of the fluctuations in the record was 'reddish', indicating the importance of non-oscillatory changes in varve thickness on a timescale of tens to hundreds of years. Overall, the power spectrum showed no statistically significant components, although one segment did produce significant peaks at 5.4 and 10.8 years. But these results were not sufficient either to support the theory that there were weather cycles in the distant past or that any such periodic behaviour might be linked to solar activity.

The conclusion has to be that, in spite of their potential, varved sediments from both recent deposits and throughout geological time provide surprisingly little evidence of significant cycles in the range from a few years to a few centuries. This can be explained in two ways. Either they mean that there are no weather cycles recorded in varves or, alternatively, varves integrate a range of meteorological variables over the year and effectively smooth out most of the cyclic behaviour in the weather. Either way, with the exception of the example we will return to in Section 4.5, there is no more than limited support for the features (e.g. 2 to 3 years, 5 to 6 years, and around 11 and 22 years) that have emerged from other studies.

4.3 Pollen records

Another feature of sediments formed in shallow lakes or bogs is the presence of pollen grains that can provide information about vegetation on the surrounding land. Where these sediments have been laid down in a regular manner, the abundance of pollen from different species of trees and shrubs provides details of climate change. Because different species have distinct climatic ranges, it is possible to interpret their relative abundance in terms of shifts in the local climate. Given that this process can occur almost anywhere on the continents, it means that pollen records have the potential to fill many of the geographical gaps which ice cores and ocean sediment records cannot cover.

[12] Crowley, Duchan & Rhi (1986).

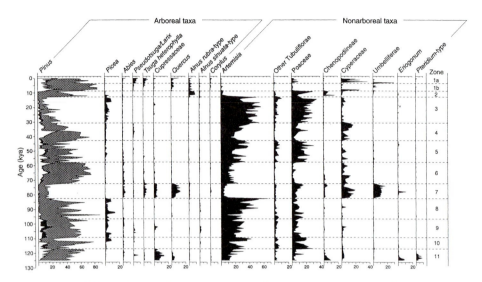

Fig. 4.6. An example of the information that can be extracted from pollen stratigraphy, using a core taken from the bed of Carp Lake in the Cascade Range of north-western USA. (With permission of McMillan Magazines Ltd.)

Most pollens (from flowering trees and plants) and spores (principally ferns and mosses) are tiny. Few exceed 100 μm (0.1 mm) in diameter and the majority are around 30 μm. It is the outer portion of the cell (*exine*) that is preserved by means of a waxy coat of material *sporopollenin*. The size and shape of this outer wall, along with the number and distribution of apertures in it, are specific to different species and can be readily identified. A typical pollen diagram (Fig. 4.6) will plot the proportion of the principal species found in a stratigraphic sequence and can be used to draw inference about how the climate has changed over the period covered by the record.

In spite of their potential for climate studies, pollen records have had little impact on the search for weather cycles. This is because the amount of pollen from different species of plants fluctuates dramatically with changing climatic conditions, and many species can completely disappear, which makes it impossible to conduct meaningful spectral analysis on the series. Where these studies have been of great value is in establishing broad global patterns of climate change. For example, recent pollen results obtained from cores drilled in Carp Lake in the Cascade Range in north-west USA (see Fig. 4.6) have confirmed that pollen records can be accurately dated back to the last interglacial, some 125 000 years ago (125 kya), and provide detailed information about how the climate has changed.[13] Furthermore, there is a close correlation between these results and those from cores from lakes in

[13] Whitlock & Bartlein (1997).

Europe, and variations observed in ice cores and ocean sediments. On even longer timescales pollen samples in ocean sediments can provide valuable independent evidence of climatic conditions on adjacent land masses (see Section 7.6).

4.4 Corals

In parts of the tropics, where there are sufficient differences in the temperature and sunshine levels at various times of the year, certain forms of coral (e.g. the star coral *Pavona clavus*) produce seasonal growth rings. The coral grows faster in warm sunny conditions than when is cooler and more overcast. The resulting bands of alternating density can be measured using X-ray techniques and used to draw conclusions about changes in seasonal water temperature. In addition, the changes in both the $^{16}O/^{18}O$ ratio and the strontium/calcium (Sr/Ca) ratio in the coral are also measures of the temperature at the time of formation. Using these techniques, measurements of temperature fluctuations in the Galapagos Islands have been obtained back to AD 1600, revealing new modes of Pacific variability. Although the main ENSO mode, centred between 4 and 5 years, was, as now (see Section 5.4), the most significant feature, other significant oscillations were found at periods of 3.3, 6, 8, 11, 17, 22 and 34 years. Both annual growth rate and the $^{16}O/^{18}O$ ratio show variance at periods equivalent to the solar and double solar periods (i.e. 11 and 22 years, respectively). In addition, the amplitude of the 11-year $^{16}O/^{18}O$ ratio cycle generally varies with the amplitude of the solar cycle, supporting previous suggestions that the solar cycle may modulate interannual to decadal climate variability in the tropics.[14]

Ancient corals have also been studied in the same way to obtain new insights into past climates. Data from around 7 to 5 kya from sites on the Orpheus Island, in the central Great Barrier Reef, the Dampier Archipelago in the eastern Indian Ocean, and Nusa Barung Island, southeast Java, in the Indonesian seaway, reveal that western Pacific surface waters were substantially warmer and saltier than present.[15] When compared with a variety of palaeoclimatic data from around the tropical Pacific basin, it becomes even more apparent that the climate of the tropical Pacific around 6 kya was distinct from that of the last several centuries. Data from both the western and eastern Pacific suggest that interannual ENSO variability, as we now know it, was substantially reduced, or perhaps even absent. The exact significance of this possibly unprecedented mode of tropical climate variability is difficult to assess because the spatial and

[14] Dunbar *et al.* (1994).   [15] Gagan *et al.* (1998).

temporal availability of palaeoclimatic data for this time period remains relatively sparse.

Corals can also record variations of freshwater run-off because they may extract and co-deposit river-borne organic compounds in their annual growth rings. The amount of organic material from the freshwater can be measured by recording the fluorescence in the yellow–green part of the visible spectrum when the coral is illuminated with ultraviolet (UV) light. Studies inside the Great Barrier Reef off the coast of Queensland, Australia,[16] have shown that the large slow-growing (5–25 mm per year) corals of the genus *Porites*, which may live for up to a 1000 years and grow to 10 m across, provide an accurate record of the amount of run-off from local rivers. This is a measure of rainfall in tropical Australia, which provides further evidence of ENSO events. This technique has also been used to study ancient coral reefs to study the climate associated with past high sea levels.

In terms of the wider climatic picture, perhaps the most important contribution that coral studies are making relates to providing information on potential links between the tropical ocean basins. Evidence of interannual and decadal fluctuations from around the Pacific and Indian Oceans, plus data from the Caribbean and Red Sea, are providing an increasingly coherent picture of how sea surface temperatures (SSTs) throughout the tropics have varied in the past and the extent to which fluctuations in different parts of the world are linked (see Section 5.7). Confirmation of the presence of a ubiquitous periodicity in the 10- to 20-year range is perhaps the most important product of these proxy records.[17] Both coral and tree-ring data confirm the existence of a coherent decadal pattern of variability around the Pacific basin, and something in this periodicity range is found in coral records in various parts of the Indian Ocean and in the Caribbean.

4.5 A cautionary tale

There is one further example of cyclic behaviour in geological records that needs to be taken on its own. This is what appeared to be the most stunning evidence of solar influence on the weather. It came from the work of George Williams in South Australia.[18] He became intrigued in 1979 by the laminated

---

[16] Isdale *et al.* (1998).

[17] There have been a large number of papers on the measurement of decadal oscillations in coral records the most relevant of which are Cole *et al.* (2000), Linsley, Wellington & Schrag (2000), Cobb, Charles & Hunter (2001), Evans *et al.* (2001), Tourre *et al.* (2001) and Cobb *et al.* (2003).

[18] Williams (1981) and (1986).

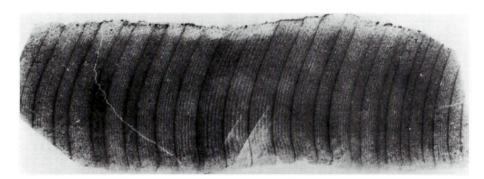

Fig. 4.7. The laminated sandstones and siltstones of the Elatina formation in South Australia suggested evidence of the Sun's influence on the climate in the Precambrian era. They show remarkably regular cyclic behaviour with the laminations coming in groups of from 10 to 14. At the beginning and end they are thin and closely spaced, while in between they are more widely spaced. (From Giovanelli, 1984.)

sandstones and siltstones in the Elatina formation in South Australia that appeared to reveal the Sun's influence on the climate in the Precambrian era. They were remarkable for their cycles – groups of from 10 to 14 laminations bordered by darker bands in which the laminations were thin and closely spaced (Fig. 4.7). Indeed, if each of these layers had been laid down in a year, they could record changes in, say, mean annual temperature or mean summer temperature. This would be truly extraordinary, as the formation would provide direct measure of cyclic behaviour 680 million years ago, when much of the planet was in the grip of a severe ice age. The geological explanation of the Elatina formation was that the layers were standard varves that were produced when turbid glacial meltwater filled a lake each summer.

The laminations studied occupied a 10-m thick unit in a 60-m thick formation. This contained roughly 19 000 laminations that contained 1580 cycles. These cycles were made up of between 8 and 16 laminations with the average number being 12. Cycles of relatively high and low amplitude tended to alternate, with the minimum lamination thickness remaining roughly constant while the maximum thickness, in the middle of each cycle, had a wide range of values. Furthermore, the thickness of the laminations and the number per cycle varied systematically. The thickest laminations tended to occur about every 26 cycles, whereas the cycles with the greatest number of laminations about every 13 cycles. The thickest laminations tended to occur in the shortest cycles.

All this bore an uncanny resemblance to the observed variations in sunspots that have occurred since 1700 (Fig. 4.8). This led to two fascinating hypotheses. First, the behaviour of the Sun had remained relatively

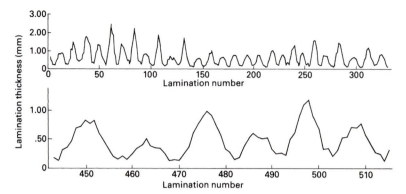

Fig. 4.8. Examples of the variations of the thickness of the laminations in the Elatina formation. The patterns observed show a cycle of about 12 laminations, with alternate cycles tending to have high and low amplitude. This pattern bears a marked similarity to the sunspot cycle (see Fig. 6.2). (From Williams, 1986. Copyright © 1986 by *Scientific American*, Inc. All rights reserved.)

unchanged over the last 700 million years. Second, for all that time, solar activity had been modulating the weather. Subsequent analysis of this record and others from Australia suggests, however, that the explanation may be lunar rather than solar.[19] The new explanation is that the laminations record the variations in sediment laid down by daily tides. So the 19 000 laminations would have been deposited in just 56 years rather than the '19 000' years originally assumed. This alternative explanation now appears to be accepted and provides only interesting evidence of the changes in the astronomical motions of the Earth and the Moon over the last 700 million years, but sadly tells us nothing about the fluctuations in the climate in the Precambrian era. Furthermore, it removes at a stroke what was widely regarded as the most convincing example of periodic solar variability affecting the weather. Without this example the evidence of periodicities in the annual layers laid down in lake beds is pretty slim.

4.6   Ice cores

A series of international programmes have resulted in an important set of climatological data being obtained from the ice caps of Greenland and Antarctica. Similar studies in major glaciers around the world have provided additional geographical coverage. Because the snow that falls each year is preserved in cold storage at the higher levels in these ice caps and

[19] Williams (1988).

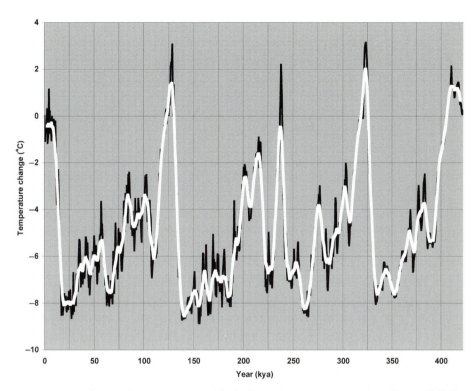

Fig. 4.9. The measurement of the hydrogen isotope ratio (H/D) in an ice core drilled at Vostok in Antarctica, showing how the ice core record extends back 420 000 years, covering the last four glacial and interglacial periods. The black line is the average values for every 500 years and the white line is the 11-point running mean of these data. (Data archived at the World Data Center for Paleoclimatology, Boulder, Colorado, USA.)

glaciers and accumulates to form ice, it preserves information on the climatic conditions when the snow fell. By extracting a core down through the ice it is possible to construct a detailed picture of climate change in the past: in the case of Antarctica over 400 kya (Fig. 4.9).[20] More importantly, in some cases it is possible to identify seasonal variations for each year over the most recent several millennia and so draw accurate conclusions about cyclic behaviour. In addition, by extending the climatic record back through the last four ice ages, this data source provides an essential link with the longer-term records that will be examined later in discussing the possible cyclic origins of ice ages.

The first ice core was drilled by the US Army in 1966 at a site in North Greenland called Camp Century. The site was chosen at a place on the ice

[20] Petit *et al.* (1999).

cap where it was estimated that the successive layers of ice would be little disturbed by the general movements of the ice as it settles and spread out. The core reached a depth of some 1400 m going right down to bedrock, and going back an estimated 150 000 years (150 kyr). This core was analysed by Willi Dansgaard and his colleagues at the University of Copenhagen, who have become leading experts at extracting climatic information from such cores.[21] Subsequently, cores have been taken from other parts of the Greenland ice cap, the most important of which were two major international projects: the Greenland Ice Sheet Project Two (GISP2),[22] which successfully completed drilling a 3053 m long ice core down to the bedrock in the Summit region of central Greenland in July 1993; and its European companion project, the Greenland Ice Core Project (GRIP),[23] which one year earlier penetrated the ice sheet to a depth of 3029 m, 30 km to the east of GISP2.

The GISP2 and GRIP records show a close similarity from the surface to a depth of 2790 m (110 kya) that provides compelling evidence of the stratigraphy of the ice being reliable and unaffected by extensive folding, intrusion, or hiatuses from the surface to this depth. This agreement between the two cores, which are separated by only 30 km (10 ice sheet thicknesses), provides strong support of climatic origin for even the minor features of the records and implies that investigations of subtle environmental signals (e.g., rapid climate change events with 1–2 year onset and termination) can be rigorously pursued. There were, however, more significant differences in the bottom 250 m of the cores that are the subject of continuing research.

Similar deeper cores have been drilled in Antarctica. The most influential results have been obtained from cores drilled at Vostok, at an altitude of 3488 m, deep in the heart of the great ice sheet of East Antarctica. The most recent core has been sampled to a depth of over 3623 m. This joint venture between Soviet, French and US workers has obtained results that extend back 420 kya (Fig. 4.9).[24] Indeed, the teams could have gone deeper, but stopped drilling for fear of punching into Lake Vostok, a huge body of water locked beneath the ice sheet that may hold more ancient organisms in cold storage. So, this core, unlike the deepest cores in Greenland, did not reach bedrock and did not have the problem of being highly compressed and possibly being distorted and sheared by

---

[21] Dansgaard *et al.* (1973) and Dansgaard *et al.* (1993).

[22] General papers on the GISP2 results include Alley *et al.* (1993), Grootes *et al.* (1993) and Taylor *et al.* (1993). Data from this ice core and the GRIP core (see 23 below) can be obtained from the World Data Center for Paleoclimatology: http://www.ngdc.noaa. gov/paleo/icecore.html.

[23] Greenland Ice Core Project (GRIP) Members (1993) and Grootes *et al.* (1993).

[24] Petit *et al.* (1999).

movement over the bedrock. So it gives a much more reliable picture of climate changes before 100 kya and is being used to provide more precise information about the cycle of glacial/interglacial fluctuations during recent ice ages.

There are a wide variety of climatic indicators that can be derived from these ice cores. These include evidence of changes in temperature, the amount of snow that fell each year, the amount of dust transported from lower latitudes, fall-out from major volcanic eruptions and the atmospheric composition of the air in bubbles trapped in the ice. Of greatest interest in the search for weather cycles is the ratio of hydrogen isotopes and the ratio of oxygen isotopes, because these can provide information about regional temperature over the entire length of the ice core. Most frequently, the ratio of oxygen isotopes (oxygen-16 and oxygen-18) has been measured, although similar information can be extracted from the ratio of hydrogen isotopes (hydrogen and deuterium). The amount of the heavier of the oxygen atoms, oxygen-18 (^{18}O), compared with the lighter, far more common isotope, oxygen-16 (^{16}O), is a measure of the temperature involved in the precipitation processes. But this is not a simple process. The snow was formed from water vapour that evaporated from oceans at lower latitudes and travelled to higher latitudes. The water molecules containing ^{16}O are lighter, and evaporate slightly more readily and are a little less likely to be precipitated on snowflakes than those containing ^{18}O. Both effects are related to the temperature, so the warmer the oceans and the warmer the air over the ice caps the higher the proportion of ^{16}O in the snow that fell. So during cold episodes in the global climate the proportion of the ^{18}O in the ice core is lower (variations in the oxygen isotope content are usually designated as $\delta^{18}O$).

Just how much information can be obtained about the variation of the ^{18}O content can be seen in the GISP2 record over the last 100 kyr (Fig. 4.10).[25] This shows that the results from Greenland contain much more detailed information on climatic variations on shorter timescales. This variability reflects the erratic nature of climate change in the northern hemisphere. In addition, the higher precipitation rates on the Greenland ice sheet enable scientists to resolve smaller time intervals and hence draw conclusions about dramatic changes occurring in just a few years.

The scale of the changes revealed in the GISP2 core, and other cores from Greenland, is remarkable[26] and was memorably described in one paper as being a 'flickering switch'. Moreover, the sudden shift to lower

---

[25] Alley *et al.* (1993), Grootes *et al.* (1993), Taylor *et al.* (1993) and www.ngdc.noaa.gov/paleo/icecore.html.

[26] See note 25.

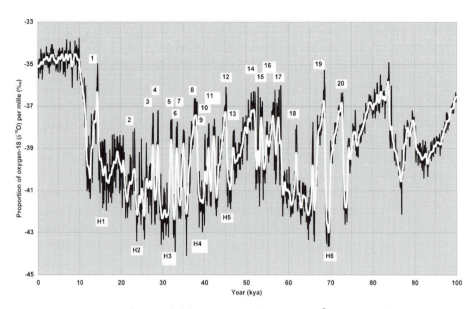

Fig. 4.10. The record of the proportion of oxygen-18 ($\delta^{18}O$) per mille (‰) in the GISP2 ice core (the black line is the data average values for every 50 years and the white line is the 41-term binomial smoothing of this data). These curves show the 20 Dansgaard/Oescheger warming events (labelled 1 to 20), six of which coincided with Heinrich events (labelled H1 to H6). The temperature range covered by these changes is reckoned to be about 20 °C between the coldest periods and the warmth of the last 10 kyr. (Data archived at the World Data Center for Paleoclimatology, Boulder, Colorado, USA.)

variability at the end of the last ice age was equally striking. The most recent 10 000 years (10 kyr) have featured a warm and remarkably stable climate, known as the *Holocene*. With the exception of a short sharp cooling around 8.2 kya, this stability is in marked contrast to what went before. Following the rapid warming from the depths of the last ice age 15 kya, there was a sharp reversal between 12.9 kya and 11.6 kya as the climate in the northern hemisphere slumped back into conditions comparable with those of the ice age. This final cold event at the end of the last ice age is known as the *Younger Dryas*, because originally it was identified in pollen records in northern Europe, notably in terms of the extent of the tundra plant *Dryas octopetela*.

Prior to the Younger Dryas, the striking feature that emerges from the Greenland ice cores is not so much the evidence of cyclic behaviour but the frequent and dramatic fluctuations throughout the last ice age. In particular, the isotopic temperature records show some 20 interstadial (known as 'Dansgaard/Oeschger (DO) events' after the scientists that first identified them:[27] the precise number is the subject of slight variations from analysis to analysis depending on whether or not different climatological groups award the accolade of being an 'event' to the most transient warmings) events between 15 and 100 kya (Fig. 4.10). These events are probably linked to changes in the rate of formation of North Atlantic Deep Water (NADW) (see Section 5.9). Typically, they start with an abrupt warming of Greenland of some 5 to 10 °C over a few decades or less. This warming is followed by a gradual cooling over several hundred years, and occasionally much longer. This cooling phase often ends with an abrupt final reduction of temperature back to cold ('stadial') conditions. The spacing between these events is most often around 1500 years, or, with decreasing probability near 3000 or 4500 years. This spacing has been linked with a 1500-year cycle (see below).

A second form of abrupt climate change is associated with what are known as *Heinrich layers* in the ocean sediments of the North Atlantic, after the scientist who first identified them.[28] These events are associated with the surging of the Laurentide ice sheet over North America that released armadas of icebergs into the North Atlantic. They have variable spacing of several thousand years (see Fig. 4.10), and left telltale signs of debris in the ocean sediments. The accompanying freshwater was apparently sufficient to shut down the formation of NADW and particularly cold stadial conditions. These cold Heinrich events appear to be associated with unusual warming in East Antarctica (see Fig. 4.9).

Turning to the search for cycles in the isotope data has the potential to provide a direct measure of the temperature change associated with past changes in the climate. Great care needs, however, to be exercised in making the conversion from isotope concentrations to actual temperatures. This is particularly so in the case of longer-term changes in the Greenland cores. Because so many features of climate change in the northern hemisphere are related to changes in circulation patterns (see Section 5.1), and, in particular, the North Atlantic Oscillation (see Section 3.7), observed changes in isotope concentrations high on the Greenland ice cap may not be a direct reflection of wider temperature fluctuations around the northern hemisphere or more specifically in the North Atlantic. While clearly the change in the temperature in these regions was a major factor in the observed isotope fluctuations, other climatic processes could be at work. For instance, shifts in storm tracks could lead to alterations in sources of precipitation and with it changes in isotope concentration that were not related to a global warming or cooling. By comparison, the symmetry of the Antarctica and

[27] Dansgaard & Oeschger (1989).
[28] Heinrich (1988), Bond *et al.* (1992) and Bond & Lotti (1995).

the circulation regime of the southern hemisphere pose less of a challenge in interpreting possible forms of climate change. For this reason, scientists are reasonably confident about being able to translate isotope variations in the Vostok ice cores into temperature changes (see Fig. 4.9).

Another potentially valuable feature of isotope data is that initially they contain complete information about the seasonal variations in effective temperature of precipitation. Newly fallen snow samples on the Greenland ice cap exhibit a seasonal variation in $\delta^{18}O$ of about 10 parts per thousand (10‰). As each year's accumulation settles, there is an exchange between the isotopic constituents of the snow and the amplitude of the seasonal cycle that falls to about 2‰. This process slows down as the air is squeezed out of the compacting snow, which after a century or two becomes impermeable ice. After that, the gradual process diffusion continues but at a much slower rate. In the case of the GISP2 core it was possible to detect annual oscillations over the most recent 15 kyr of the record. The much slower rate of accumulation of snow in Antarctica, especially at the intensely cold remote site of Vostok, means that there is no scope for analysing annual rings over any substantial length of time.

Even in Greenland, where there is a clear annual cycle, it does not mean that the seasonal cycle is locked in with such precision that conclusions can be drawn about the weather for each year over this lengthy period. Because exceptionally snowy years or the formation of ice after rare summer thaws can alter the rate of diffusion, there are occasional bands of excessive annual oscillation deep in the ice. Nevertheless, the variations between warmer and colder periods on the timescale of a decade or longer are permanently locked into the ice cores, and in the case of the Camp Century core some profoundly influential inferences were drawn about shorter variations during the most recent 800 years (Fig. 4.11a) (see Section 8.2). Spectral analysis of section of the core taking 10- to 20-year segments showed a considerable number of features (Fig. 4.11b) with the most significant being at 78 and 181 years.[29] Studies of the rest of the core produced a confusing mixture of periodicities, some of which appeared to support the case for longer-term cycles (e.g. 350- and 2500-year periodicities). These results have now been overtaken by the results of the GISP2 and GRIP cores.

Because the isotope values recorded in the ice core are the complex product of both precipitation processes and diffusion during settlement, great care must be exercised in analysing annual figures. Using a single record is fraught with difficulties. Studies of various shallow cores in Greenland have shown that about half the variation recorded was of non-climatic

[29] Dansgaard *et al.* (1973).

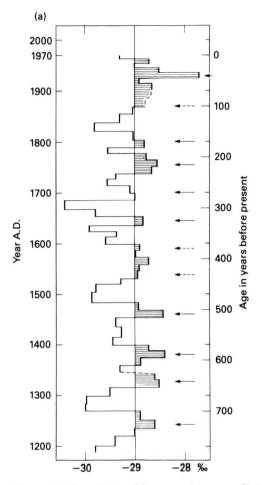

Fig. 4.11. (a) The variation of the oxygen isotope profile in the Camp Century ice core during the last 800 years, and (b) its power spectrum, showing the two principal periodicities at 78 and 181 years. (From Dansgaard *et al.*, 1973.)

origin due to local effects depending on the rate of deposition of the snow. Moreover, the records exhibit the properties of 'red' noise (see Section 2.7), which complicates matters further. These effects tend to be ironed out over the longer term, but they make the use of annual records risky. To get round these problems, workers from University of Copenhagen and the US Army Cold Regions Research and Engineering Laboratory have combined observations from three ice cores taken from different places on the Greenland ice cap.[30] Because the origin of the precipitation at these sites was different,

[30] Hibler & Johnson (1972).

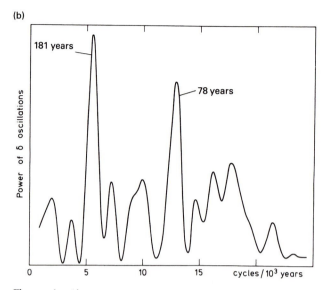

(b)

Fig. 4.11. (*cont.*)

the isotope figures tended to iron out the non-climatic factors. In this way it was possible to form a more accurate set of annual values for the period 1244 to 1971. Using two-year non-overlapping observations and MESA techniques, they produced clear evidence of a cycle of 20 ± 0.5 years, which was significant at the 99% level. So the Greenland ice-core data add to the general body of evidence for the ubiquitous cycle with a period of around 20 years. In the GISP2 core, the Holocene is characterised by fluctuations around a long-term stable mean dominated by a 6.3-year oscillation.[31] This period is close to the less convincing periodicity in the North Atlantic Oscillation (see Section 3.7). Other notable periodicities identified in this study include 11 and 210 years, which are also found in the solar-modulated ^{10}Be and ^{14}C records (see Section 6.1). This core does not, however, provide any support for the 80- and 180-year cycles in the orginal Camp Century core that exerted considerable influence on thinking about global warming in the 1970s (see Section 8.2).

Even more striking evidence of a solar signal comes from analysis of dust levels in the GISP2 core. Using a laser light scattering technique, Michael Ram and colleagues at the University of Buffalo, Buffalo, New York, have produced clear evidence of a solar signal in the variations in the dust content in the ice core.[32] The most impressive results came from the bubble-free ice below a depth of 1660 m and gave continuous annual record of dust

[31] Grootes & Stuiver (1997).   [32] Ram, Stolz & Koenig (1997) and Ram & Stolz (1999).

concentration from 11 kya to at least 125 kya. Throughout this period, there was repeated evidence of an 11-year cycle, the period of which typically fluctuated between 7 and 17 years. This behaviour tallies closely with that of the sunspot cycle (see Section 6.1). In addition, there is evidence of cycles centred broadly on 90 and 200 years, together with the 11-year cycle behaving in a way that is strongly suggestive of a 22-year cycle. Working on the basis that the amount of dust, which peaks each spring, is a measure of continental aridity and hence of large-scale precipitation patterns, this is fascinating evidence of solar influence on the global climate (see Section 6.3).

When it comes to evidence of longer cycles, the results from these cores are mixed. The most intriguing result is a highly significant quasiperiodic feature in the GISP2 core analysis with a period of around 1500 years, give or take a few hundred years.[33] As noted above, during the last ice age it is closely linked with the somewhat erratic occurrence of DO events but it has persisted through the Holocene. Furthermore, there is increasing evidence of it in ice-rafted debris in the North Atlantic, peatlands in western Canada, and the eastern winter monsoon in China.[34] But there are a number feature of these results that have led to a spirited debate about the significance and cause of this quasi-cycle. First, its period does vary appreciably both within studies and from place to place. Second, results from the GRIP raise a number of challenges. Any link with DO events has to explain why, although the most frequent spacing is around 1500 years, there are longer gaps, which appear to be a multiple of this period. Then there are interesting physical questions to be answered about the sudden nature of these events, which smacks of chaotic behaviour (see Section 8.1). The answer may lie in the even more complex behaviour known as *stochastic resonance* (see Section 7.7).

Turning back to Antarctica, in terms of detecting shorter-term cycles, the data are less rewarding. An early ice core was drilled at Byrd station in the late 1960s. Attempts to date even recent accumulations came up with widely fluctuating values varying from 8 to 70 centimetres of ice a year. These figures were too irregular to justify dating the Byrd core by means of annual layers. Moreover, because of the more complicated glaciological conditions in the vicinity of the site it was difficult to provide an absolute dating of the core. While the broad pattern of isotope fluctuations showed the same ice age pattern, regular cyclic fluctuations were evident. When the general features of the Byrd ice core over the last 80 kyr or so were matched up with those of the Camp Century core, it was possible to detect a

[33] Bond *et al.* (1997).
[34] See Bond *et al.* (1997) and (2001), Campbell *et al.* (2000) and Lu *et al.* (2000).

2400-year oscillation back to 20 kya. Otherwise, the core provided little evidence of cyclic behaviour.

The cores from Vostok move us into a new scale of climatic exploration, as they make a major additional contribution to understanding the dynamics of ice ages. This is because, unlike the Greenland cores, they provide reliable data (see Fig. 4.9) covering the last four ice ages.[35] These cores have proved to be a treasure house of climatic information. Of particular interest are the oxygen and hydrogen isotope measurements, plus those of carbon dioxide and methane trapped in the tiny air bubbles in the ice. The Franco-Soviet group argues that measurements of the variations in the abundance of the heavy hydrogen isotope deuterium provide a slightly better estimate of past temperatures. The significance of the changes in the carbon dioxide and methane concentrations, which show a close parallelism with the inferred temperature profiles, will be discussed in Chapter 7.

The deuterium content of the Vostok cores was measured for 100-year intervals over their entire length. The timescale of the core was defined in terms of the glaciological analysis of the region of the East Antarctic ice cap beneath the Vostok station. Various forms of spectral analysis have been used to examine the time series generated from the cores. In all these studies, three major periodic features are found, at around 110, 42.5 and 23.3 kyr. These cycles constitute around half the total variance in the series. Furthermore, they closely match the principal periodicities in the Earth's orbital parameters, and tie in well with the observed periodicities in deep ocean sediment cores (see Section 4.8), which provide the clearest possible evidence of cyclic behaviour in the climate. To examine these observations we now need to change tack. The analysis of the Vostok ice cores has shifted our timescale from that of years, decades and centuries to that of tens of millennia. In addition, we must address the evidence of the most important long-term shifts in the climate, namely the ice ages. But before doing so there is one other aspect of the cryosphere to consider, which goes some way to filling in the gap between periods of a few centuries and tens of millennia – the movement of glaciers.

4.7   Glaciers

Alongside ice-core data, evidence about the extent of glaciers is an important source of information about past climate change. As a consequence, a huge amount of work has been done around the world to measure the

---

[35] Petit *et al.* (1999).

expansion and contraction of glaciers. These measurements consist mainly of dating the age of terminal moraines left by glaciers when they expanded during cold, wet climatic intervals such as the Little Ice Age between 1550 and 1850. Because the movement of glaciers is an integrated response to changes in the weather over a number of years, such measurements are a useful guide to climate fluctuations on a timescale of decades to centuries to a few millennia.[36]

There are, however, challenges in using glacier measurements to obtain evidence of climatic cycles. Because glaciers remove all evidence of previous movements on the ground they cover, the terminal moraines are records of their greatest extent. This means that intermediate less extensive surges may have been scrubbed from the record. As a consequence, they provide a more detailed picture of recent fluctuations while in the more distant past only the greatest changes are preserved. So the records tend to show shorter-period fluctuations in the last millennium or so, and rather longer periodicities prior to then (Fig. 4.12). Their expansion and contraction appears to be linked to level of solar activity as inferred from radiocarbon (^{14}C) levels in tree rings (Section 6.1).

In spite of these limitations, glacier studies are important for three reasons. First, they are a major source of information about climatic fluctuations since the last ice age. Second, because they are found in many parts of the world, it is possible to confirm that the most significant changes have been synchronous on a global scale. Third, they do suggest that the changes have been quasi-cyclic, and some workers have sought to prove more regular variations. In particular, George Denton of the University of Maine has compiled considerable evidence of a 2.5-kyr periodicity in global climate.[37] Subsequently, evidence of more frequent millennial scale fluctuations in both glacier records (see Fig. 4.12) and elsewhere in the climate system have been linked with the periodicity of around 1.5 kyr.

4.8   Ice ages and ocean sediments

The huge range of evidence on the ice ages means that instead of looking for the cycles in individual types of proxy data, we now need to examine how all the evidence has been combined to produce the most convincing case for cyclic behaviour in the climate. In so doing, we will, however, give

[36] See for example Denton & Karlén (1973), Neftel, Oeschger & Suess (1981) and Grove (1988), Chapter 10.
[37] Denton & Karlén (1973).

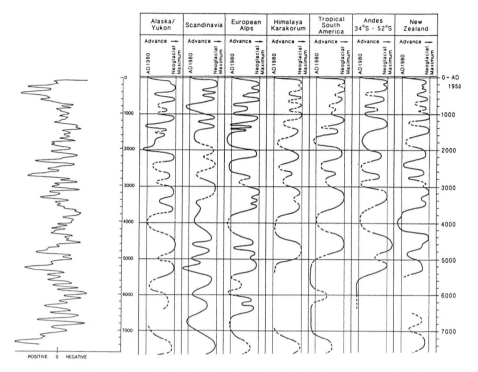

Fig. 4.12. An analysis of the fluctuations of glaciers in the northern and southern hemispheres during the last 7600 years, compared with radiocarbon production variations. Periods of negative radiocarbon production signal lower solar activity and are associated with glacial advance, which is a sign of a cooling climate. (Neftel Oeschger & Suess, 1981, Berger, 1990.)

pride of place to ocean sediments because of the pivotal role they played in establishing this understanding. The work surrounds the investigation of the frequency of ice ages over the last one to two million years. It requires us to address a whole body of data assembled to provide the chronology of these major climatic events. The work is now dominated by the results of measuring ocean sediments, but there is a long history to our emerging knowledge of the nature and scale of ice ages. Until the technology of obtaining cores from the bottom of the deep oceans was developed in the 1950s and 1960s, we had to rely on land-based observations. To understand how our thinking is built on a range of other observations we need to look more closely at the history of thinking about ice ages.

Throughout the nineteenth century, following the observations of James Hutton and the major work of Louis Assigiz, an apparently consistent body of geological evidence was collected to provide a clear-cut explanation of past ice ages. This view was encapsulated in the work of

Fig. 4.13. One of the protozoan foraminifera, which make up part of the foraminiferal ooze in the deep-ocean sediments. Analysis of the species and isotope content of the skeletal remains can provide valuable information about past climatic change. (From Lambert, 1988.)

Penck and Bruckner published in 1909. Stated simply, this indicated that over the last million years there had been four cold glacial periods of duration about 100 000 years, separated by warm interglacial periods ranging from 125 000 to 275 000 years in length. The present warm period had started about 25 000 years ago and was destined to last indefinitely.

This orderly view held sway for more 50 years, and is still to be found in some relatively modern texts on the subject. But, in the 1950s a new picture began to emerge. This came from a set of papers, published by Caesari Emiliani, when at the University of Chicago,[38] on the oxygen isotope ratios of the fossil shells of pelagic foraminiferal species (Fig. 4.13) found in the Caribbean and equatorial Atlantic deep-sea cores. A major component of the deep-ocean sediments is the shells of these tiny plankton species that live in the surface waters. Because of biological processes, in forming the calcium carbonate that makes up their shells these creatures absorb more ^{18}O at lower temperatures. When they die, they rain down to the ocean bottom and very slowly build up the sediment. By collecting shells belonging to identifiable species known to have lived in surface waters within various temperature ranges it is possible to build up a chronology of temperature changes at a given location. But because of the slow rate of accumulation, often no more than a metre every 100 000 years, short-term changes of less than a thousand years or so are blurred out because bottom-dwelling creatures thrive on eating mud and churn up the top layer. These records are, however, ideal for examining long-term fluctuations, and deep-sea cores have been obtained from all around the world to provide a detailed picture of the geographical variations of the climate changes that occurred during the ice ages.

Over the years a huge range of analytic techniques have been developed to tease out climatic information from ocean sediments. Apart from the types of foraminifera found in the sediments and their isotope ratios, other measurements include the ratios of various elements in their

[38] Emiliani (1955).

skeletons (e.g. Sr/Ca, Cd/Ca and Mg/Ca), which provide information about not only SSTs, but also nutrient levels in the water. Then the levels of certain types of organic chemicals, known as *alkenones*, formed from the decay of dead algae that lived in the surface waters, are an independent measure of SSTs. In addition, the size distribution of sand grains in sediments formed below ocean currents can provide a measure of the strength of these currents and, hence, a picture of climatic conditions in the past. Here, however, we will concentrate on the early work, starting with that of Emiliani, that established the basis for the whole body of evidence that supports the case for the orbital theory of cause of ice ages.

Emiliani published a series of papers in the late 1950s, which showed clearly that there had been seven ice ages in the last 700 000 years, and that they occurred every 100 000 years or so. But it was another decade or more before the established geological view was overturned. Furthermore, the analysis of deep-sea cores had to take into account another factor in measuring ^{18}O: in the build-up of the ice sheets during the ice ages, ^{18}O accumulates in the oceans. This is because water molecules containing heavy oxygen evaporate more sluggishly than ordinary water. So the heavier water molecules are less likely to end up in the ice sheets. If the ice sheets grow to contain a measurable proportion of the oceans, this can be detected as an increase in the ^{18}O concentration in the foraminifera in ocean sediments, so the isotope measurements contain information about SSTs and ice-sheet accumulation, and, as in the ice-core data, variations in the oxygen isotope content ($\delta^{18}O$) provide a measure of past temperature changes.

To unscramble these complementary effects, Nicholas Shackleton at Cambridge University analysed the $\delta^{18}O$ of the fossils of these creatures that lived on the ocean bed.[39] Here, the temperature changes have been negligible. The results showed that it was possible to build up an unambiguous picture of the variation of ice volume during the ice age cycles of the last million years so. When combined with other studies of the distribution of different types of surface plankton over time, together with more convincing dating techniques and supporting evidence from land-based data, an unequivocal picture of past climate change has emerged. But the foundation of this new view of the ice ages has been the large number of deep-sea cores, principally obtained as part of the CLIMAP programme (a joint project of four US universities). This ambitious programme aimed not only to establish unequivocal evidence of past climate changes but also to map out in more detail the climate at certain specific times of the year. These 'snapshots' of the past global climate include analysis of the

[39] Shackleton & Opdyke (1973) and CLIMAP (1976).

post-glacial climatic optimum of 6 kya, the nadir of the last ice age at 18 kya and the height of the previous interglacial around 120 kya.

The important dating marker was to identify a magnetic marker in the deep-sea cores. This reference point is based on the fact that from time to time in the Earth's history the magnetic field of the planet reverses. The last occasion was 700 kya and the reversal left an indelible mark in the sedimentary layers. So, while there had been doubt about the dating based on assumed rates of sedimentation, there could be no doubt about the presence of this magnetic horizon. This meant that the oscillations first observed by Emiliani and found in so many other ocean cores could be dated with precision.[40]

The supporting land-based data came from parts Central Europe. In places where, unlike many parts of northern Europe, the land had never been covered with ice, it had built up a complete sequence of wind-blown soil deposits. They also showed more frequent climatic variations with eight major shifts since the magnetic reversal of 700 kya. In contrast, areas to the north, where much of the work was done on establishing the chronologies of the ice ages, had effectively had the slate wiped clean by glacial action. But because the evidence of alpine moraines and glacial deposits had been interpreted in a particular way, it required the combinations of the deep-sea cores, the magnetic dating and the evidence of the same frequent climatic fluctuations in the soils of Central Europe to bring about a fundamental review of the established view.[41]

The dramatic message, at the time, from the deep-sea cores was that beyond any shadow of reasonable doubt the climate of the last million years has been dominated by three major cycles of periods around 21, 41 and 100 kyr. Subsequent work has led to the stacking of many open-ocean sediment records to produce a standard curve for changes in $\delta^{18}O$ over the last 600 kyr known as SPECMAP (standing for spectral mapping of climatic parameters),[42] which show close parallels with the ice core results from Vostok (see Fig. 4.14), and has confirmed these broad features of the climate record. The various aspects of the debate about whether the observed behaviour can be explained by astronomical theories will be explored later in the book. Much depends on detailed aspects of the accurate dating of both ocean sediments and other proxy records, notably *speleothem* observations made in the United States (see Section 4.9). Furthermore, as is evident in

[40] A succinct review of these developments appears in Hays, Imbrie & Shackleton (1976).

[41] Hays, Imbrie & Shackleton (1976).

[42] SPECMAP No I Archive: http://www.ngdc.noaa.gov/mgg/geology/specmap.html.

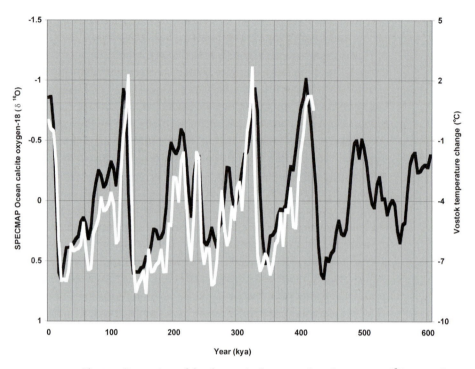

Fig. 4.14. Comparison of the changes in the proportion of oxygen-18 ($\delta^{18}O$) per mille (‰) in the foraminifera sampled in the SPECMAP ocean sediment analysis (the black line) and the temperature change inferred from the ice core drilled at the Vostok research station in Antarctica (the white line) (see Fig. 4.9). (Data archived at the World Data Center for Paleoclimatology, Boulder, Colorado, USA.)

Fig. 4.14, there are interesting differences between the rhythms identified in the ocean sediments and those found in the Vostok ice cores.

To understand how the differences arise and how they may be reconciled, we need to consider both possible explanations of the cycles and the dating techniques used to define the age profiles for the various types of cores. These cycles correspond closely with the variations in the Earth's orbital parameters that will be examined in detail in Chapter 6. Here, it is important to note that not only was the evidence of the proxy data unequivocal but also there was a well-rehearsed physical theory to explain the observed changes. This had first been proposed by the Scot James Croll in 1864, but is usually attributed to the Serbian geophysicist Milutin Milankovitch. It was he who developed the theory of how changes in the eccentricity of the Earth's orbit, together with the precession of the equinoxes and variations of the tilt of the Earth's axis, could lead to changes in the amount of solar energy at different latitudes and different seasons. These long-term changes, it was argued, could trigger the expansion and contraction of the ice ages.

The sudden acceptance of the new ice age chronology in the early 1970s, together with the existence of a plausible physical explanation, led to a surge of work to refine the climatic models of the ice ages. Within a few years, work around the world, and in particular by members of the CLIMAP team, had established the Milankovitch theory as part of climatic orthodoxy. The importance of this conversion is not just that it marked a major step forward in our understanding of the Earth's climate but also that it sets standards by which other proposed cycles must be judged. In the case of the Milankovitch theory the combination of good statistical evidence with a plausible physical mechanism to explain the observed changes was what was needed to establish scientific acceptance. Although there are details of the theory which are the subject of considerable debate (see Sections 6.4 and 7.5), including lingering doubts about the timing and scale of various events around the world (see Section 4.9), it still stands head and shoulders above other proposed climatic cycles. This confirms the observation in the introduction to this chapter that the ice ages provide a good yardstick against which to gauge other 'cycles'.

This having been said, it is important to note that such is the confidence in the evidence of the orbital cycles that they have been used to 'tune up' the dating of ocean sediments.[43] Because the rate of sedimentation may have varied over time, there is a problem with dating the sediments on the basis that they were deposited at a constant rate, as it may lead to significant errors. Moreover, sedimentation rates will vary from place to place around the world. But as more and more cores were examined a consistent picture emerged. Not only were the broad features of the major fluctuations attributable to orbital forcing easily identified but also a sequence of less dramatic features was found in many of the records. This analysis did not rely solely on the oxygen isotope variations in the shells of deep-water plankton: other measures included the relative abundance of surface-water plankton (a measure of summer SSTs) to provide a cross-check on any adjustments to the timescale. It also used four different approaches to link the observed changes to the calculated latitudinal variations in solar intensity that are the 'pacemaker' driving the climatic ups and downs (see Section 6.4).

The conclusion of this tuning up process is to produce improved dating of major climatic events over the last 300 kyr.[44] The accuracy of this dating is about ±5 kyr. So the metronome of orbital variations can be used to provide a better picture of the timing of past climatic fluctuations. This approach must be treated with care, however, when considering the amount of variance in the climatic record that can be accounted for by these orbital

[43] Martinson *et al.* (1987).   [44] Martinson *et al.* (1987).

cycles. There is also an element of trying to lift oneself by one's own boot-straps when the sediment timescale is used to calibrate the ice core obser-vations from Vostok, as the glaciological calculations used to construct the timescale for these cores has the same inherent limitations as the sediment figures because of assumptions about both precipitation rates and the flow characteristics of the ice sheet in the vicinity of Vostok station. Nevertheless, the correspondence between the results from two very different sources is striking (see Fig. 4.14).

As for the matter of the variance in the sediment records, there is no doubt that the major features in the climate record can be attributed to the forcing effects of the orbital variations, although there is still a consider-able amount of unexplained variance.[45] Indeed, while fully 80% of the $\delta^{18}O$ variance is distributed in the frequency range of the two components of the orbital forcing used to tune up the record, only 25% of the total variance of the record can be described as a linear response to this forcing. Invoking the 100-kyr eccentricity to explain the ^{18}O variations increases this figure by about 50%. Moreover, 72% of the variance in the frequency range of the eccentricity can be ascribed to this orbital variation.

These results show that in the longer term the large-scale fluctuations in the global climate can in part be attributed to periodic variations in the Earth's orbit. However, about half the variance remains unexplained. This means that while the Milankovitch theory of the ice ages is by far and away the most impressive case of periodic behaviour of the climate, the underlying natural variability of the climate remains. This, together with the inevitable limitations in the data, leaves considerable uncertainty about how the orbital variations produced the observed climatic changes. Most of this debate focuses on the fundamental issue of how relatively small changes in insolation at different latitudes at different times of the year can produce such large changes in the climate; focusing particularly on how the role of the 100-kyr eccentricity can have so much impact.[46] In addition, the whole question of the estimated timing of the various glacial and interglacial events has remained a ticklish issue. In particular, some land-based data (see Section 4.9) have been a thorn in the side of the proponents of the Milankovitch theory. The question of how various physical processes can produce the observed climate change, and the current state of the debate on timescales, will be explored in Chapter 6.

Because ocean sediment records are dominated by the cycles in the Earth's orbit, less attention was paid to shorter periodicities during the 1970s and 1980s. This reflected the relative lack of evidence of cycles between

45 Hays, Imbrie & Shackleton (1976).
46 Imbrie & Imbrie (1980) and Imbrie *et al.* (1992) and (1993a).

1 kyr and 10 kyr, and also that there was no well-established physical case for periodicities in this range. But, in recent years, stimulated in a large part by the results from the latest Greenland ice cores (Section 4.6), there has been a surge of interest in this area.[47] This picked up on earlier evidence of cycles in this range, notably at periodicities of 2.5, 4.7 and 10.3 kyr,[48] and has concentrated on two principal areas: the widely observed 1450-yr periodicity (see Sections 4.6 and 4.7) and a less-well-established feature at around 5 kyr. The fact that these are frequently referred to as 'quasi-cycles' and are usually quoted with rather wide bandwidths (e.g. 1450 ± 500 years) is a measure of the uncertainties surrounding these features. How these quasi-cycles could be the product of the non-linear response of the climate to orbital cycles, or linked in some way to other external or internal periodic or quasiperiodic features of the climate, will be discussed in Chapters 6 and 7. In addition, data obtained from a North Atlantic deep-sea sediment core covering the last 11.5 kyr shows sediment colour variations, a proxy for changes in NADW circulation, with periodicities of 550 and 1000 years, as well as the 1450-year periodicity.[49]

Another feature of the longer-term changes observed in both ocean sediment and ice core data for the period covering the last glacial shows the striking suddenness of some of the changes that occurred. In particular, six of the most dramatic warmings over the period 16 kya and 70 kya, which can be seen in the Greenland ice core (see Fig. 4.10), follow hard on the heels of Heinrich layers (see Section 4.6) in the ocean sediment data.[50] These layers are thought to been produced by debris carried out into the North Atlantic by a surge of icebergs resulting from the sudden collapse of part of the ice sheet covering North America. In addition, the more frequent Dansgaard/Oeschger events (see Section 4.6), which may be related in some way to the 1450-year cycle, are another striking feature in the ocean sediment records. These sudden changes raise important questions about both the nature of the deepwater circulation of the North Atlantic and how influxes of fresh water might have altered the deepwater circulation of the North Atlantic (see Sections 5.9 and 7.3).

4.9   Other proxy measurements

Alongside the principal proxy measurements described above, there are wide ranges of other sources of information, which are used where the circumstances permit. Wherever a natural process lays down long-term

[47] Bond *et al.* (1997).   [48] Pestiaux *et al.* (1988).

[49] Chapman & Shackleton (2000).   [50] Dansgaard *et al.* (1993).

deposits of material at an approximately constant rate, there is the prospect that some climatic information will be recorded. For instance, the deposition of calcium carbonate encrustations by running water in caves (*speleothems*) records changes in the isotopic ratios in precipitation. So, where such encrustations build up over a long time, they have the potential to provide useful climatic information. One of the best known examples of this process is a 36-cm layer that built up in Devils Hole, Nevada, USA, which has provided valuable data for the period from 500 kya to 60 kya, after which the encrustation ceased.[51] The importance of this particular work is that it provided the most substantial challenge to Milankovitch orthodoxy. Both the timing and nature of the observations of the impact of the last five glacial periods appeared to be seriously at odds with the orbital theory of the ice ages. The state of play on this continuing debate is reviewed in Section 7.6.

Speleothems have also supplied information on more recent climatic fluctuations. Studies in south-west Ireland have produced detailed information about changes in the oxygen isotope ratio of rainfall over the last 10 000 years.[52] Apart from providing independent evidence of the more dramatic shifts in the climate of the North Atlantic, this work has produced evidence of periodicities of 78, 169 and 625 years. Similar work on a formation of stalagmites in northern Oman has identified a strong coherence between solar variability and the monsoon during the period 9 to 6 kya.[53] Statistically significant periodicities were found at 87, 134, 205 and 779 years. Where greater rainfall produced more rapid build up of calcite, higher resolution studies revealed periodicities of 3, 7.5 and 24 years.

On shorter timescales, stalagmite studies relating to rainfall during the last 1000 years near Beijing, China,[54] provide a useful check with historic records (see Section 3.4). A wavelet spectral analysis of stalagmite annual lamina thickness found that the dominant cycles were 2, 3.3, 5–6, 10–12, 14–18, 133 and 194 years. It also found that some of the cycles were dominant at one time and others at another, and within the decadal and interdecadal bands the period moved about with time. Usually, any dominant cycles were stronger in wet periods when the laminae were thicker. In dry periods, the laminae were thinner and the power of the principal cycles was also weaker. So, as with many other proxy records, the power and frequency of the dominant cycles changed over time, which is also consistent

[51] Winograd *et al.* (1988), (1992) and (1997).
[52] McDermott, Mattey & Hawkesworth (2001).
[53] Neff *et al.* (2001).   [54] Oix *et al.* (1999).

with the equivocal picture obtained from the Chinese written records (see Section 3.4).

4.10   Predators and prey

Although not a source of information on weather cycles, efforts to link fluctuations in the population of certain animals to sunspots provides a good example of how fluctuations in natural phenomena can be the source of diverse forms of interpretation. It is also a fascinating example of how complex non-linear (*chaotic*) systems can produce apparently convincing example of 'cycles'. The best-known case of this behaviour is the predator–prey relationship of the Canadian lynx (*Lynx canadensis*) and the snowshoe hare (*Lepus americanus*), originally found in the trapping records of the Hudson's Bay Company.[55] While available hare statistics show a marked periodicity of around 10 years, the number of lynx furs traded has a quite extraordinary regularity extending over more than 11 cycles (Fig. 4.15). This has led to lengthy speculation that these fluctuations were linked to the climate and hence to sunspots ever since Charles Elton's paper in 1924 on periodic fluctuations in animal populations.[56] Although Elton later abandoned the causal climatic link between fluctuations in sunspot numbers and animal populations, the proposal continued to excite interest. A recent study[57] shows beyond reasonable doubt that the observed fluctuations cannot be linked to sunspots.

The importance of these relationships lies in how ecological models can show that populations of animals can exhibit both periodic and chaotic behaviour. In the case of predators and prey these effects can be amplified, as with the snowshoe hare and the lynx. Here, the underlying periodicity appears to be driven by the dynamics between the hare and its food supply, while the lynx, at the top of the food chain, is even more controlled by these periodic interactions. Such a dramatic example of how apparently relatively simple non-linear system can produce such dramatic periodic effects is a timely reminder of how careful we need to be when ascribing causes to observed quasiperiodic climatic behaviour. In an immensely complicated non-linear system like the global climate we should never underestimate its capacity to generate its own forms of apparently quasiperiodic behaviour.

[55] Comprehensive details of Lynx populations can be found on the Global Population Dynamics Database website: http://cpbnts1.bio.ic.ac.uk/gpdd/.
[56] Elton (1924).   [57] Lindstrom (1997).

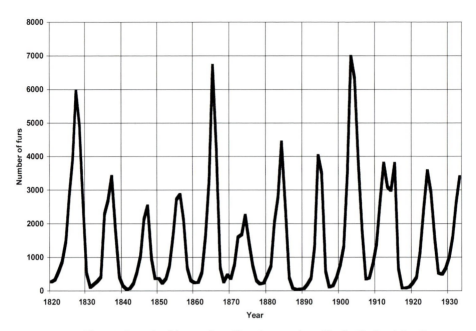

Fig. 4.15. Records of the number of lynx furs purchased by the Hudson's Bay Company in the McKenzie River region provide a graphic measure of the periodic fluctuations in the population of this species. (Data from the Global Populations Dynamics Database, University College, London, England.)

4.11 Economic series

Finally, there is one other form of proxy data that has been widely used to make inferences about past climatic conditions. This is economic series (e.g. cereal prices and wine harvest dates). These present even more challenges when looking for evidence of weather cycles. The fact that the effects of weather have been combined with factors relating to demography, market forces and social forces may make the data impenetrable. There are, however, two important reasons for conducting a brief survey of the subject. First, as noted in Chapter 1, the existence of lengthy economic series has proved a rich seam in the search for cycles. Second, these series provide some indication of the potential pay-off for identifying real cycles.

One of the most interesting series is the trend-free index of European wheat prices from 1500 to 1869 prepared by William Beveridge.[58] Better known as the author of the report that formed the blueprint for the welfare state in post-war Britain, Beveridge regarded his work on price series in the 1920s while Principal of the London School of Economics as

[58] Beveridge (1921).

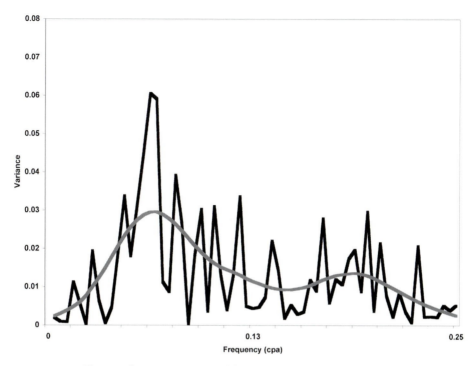

Fig. 4.16. The power spectrum of the Beveridge wheat-price index for the period 1550 to 1805, showing that a wide range of periodicities are present, calculated using data from Beveridge (1921).

his most important research work. Analysis of the harmonic components of the European wheat series produces a complex power spectrum (Fig. 4.16) with a broad peak having a periodicity of about 16 years. But the complexity of the spectrum is a clear indication of the futility of using this analysis to produce forecasts of subsequent price movements. Furthermore, the extent to which these fluctuations are the consequence of climatic events, as opposed to other developments that mapped out the fabric of European society throughout two and a half momentous centuries, is far from clear.

The record of wine harvest dates provides a more promising source of climatic information. Because the date that the grapes reach maturity is closely controlled by the weather during the growing season, it provides a useful proxy for temperature. Cool springs and summers produce late harvests while warm ones have the opposite effect. Moreover, because of the economic importance of the harvest, wine-producing areas in France and adjacent countries have a long tradition of keeping accurate records on when the grapes were ready for picking. This enabled Emile Ladurie and colleagues in France to produce a record of average wine harvest dates in northern France, Switzerland and the Rhine valley that stretch from 1482 to 1879.[59] This series provides an unrivalled source of information about the temperature in the summer half of the year for north-west Europe.

This record contains little evidence of periodic behaviour, although there is a 7-year feature that appears in the second half of the sixteenth century and during the seventeenth century. Furthermore, the power spectrum shows no obvious evidence of being 'red'. Moreover, on closer examination, some of the longer-term variations do not tally with known temperature trends. An example of this is the apparently cooler period during the second half of the eighteenth century, which was not confirmed by the limited available thermometer records. The reason for this discrepancy was that sweeter fuller wines became more fashionable in French society. As a consequence, the wine producers chose to allow the grapes to ripen longer on the vine before picking. So the later wine harvest dates reflect the shift in the market for wines, not the changing climate.

An even more dramatic example of these problems is the immigration figures for the United States. The figures, which show a marked 18-year cycle imposed upon a broad peak at the beginning of the twentieth century, have been cited[60] as evidence of the economic activity in the United States being affected by the drought cycle in the central plains (Section 4.1). The marked troughs in the immigration figures are centred on 1860, 1877, 1898, 1918 and 1934. While the second and the third may conceivably reflect in part the economic effects of the drought in the Mid-West, the others have to be attributed to other causes. In particular, the fourth and fifth were largely the product of the First World War and the worldwide economic depression respectively. It is difficult to argue that these two global cataclysms were primarily the consequence of rainfall variations in the Mid-West United States, although the severe drought of the 'Dust Bowl' years of 1934 and 1936 may have compounded the US economic difficulties and put an additional brake on immigration. A similar argument can apply in the case of the first trough: at the time the economy of the United States was dominated by the activities of the East Coast and the southern states of the Confederacy, so the impact of rainfall variations further west was inevitably much smaller.

The conclusion that must be drawn from this brief review is that the evidence of weather cycles in economic series is at best faint. There is no disputing that these series often show marked periodic or quasi-periodic behaviour. But linking these oscillations in even the most weather-sensitive series with meteorological cycles is exceedingly difficult. So while there

[59] Le Roy Ladurie & Baulant (1980).   [60] Burroughs (1997), p. 157.

Table 4.1. *A summary of the most significant periodicities in proxy data (excluding the QBO and orbital variations)*

Source	Period (years)																		
US drought index (tree rings)						18.6	22			90									
Finnish tree rings							23	30		90									
Fennoscandian tree rings				11				31–34		86–100									
Bristlecone pines											114			208		667		1400	
Galapagos corals	3	6	8	11	17		22	34											
Geological sediments		5.6		11			22			90?				200?	400?				
Greenland ice cores		6.3		11			20–22			90				200				1470	
Antarctic ice cores																550	1000	1500	2500
Ocean sediments	3.3	5–6		10–12	14–18														
Chinese speleothem												133		194					
Irish speleothem									78				169			625			
Oman speleothem	3		7.5				24			87		134		205		779			

is no dispute that about the fact that weather extremes have immediate economic effects, these are combined with many other factors in even the most basic economic series. What is more significant is that some of these series, like many other economic series, do show such a propensity to quasi-cyclic behaviour. Just as in the case of wildlife series (Section 4.10), this may tell us more about complex non-linear systems than it does about the links between the weather and economic activity.

4.12 Summary

A variety of conclusions can be drawn from this review of the evidence of cycles in proxy data. First, in the case of periodicities from a couple of years to a few decades the same picture emerges as from the instrumental records reviewed in Chapter 3. This is the frequent observation of an approximately biennial periodicity that may well be linked to the QBO in the stratosphere, plus the regular appearance of vague cycles around 3 to 4 years, possibly linked to ENSO fluctuations, and 5 to 7 years. In addition, there is some support for the 11-year cycle, and more substantial evidence for the 20-year cycle, although in general it is not possible to distinguish between this being the 18.6-year lunar cycle or the 22-year double sunspot cycle (see Sections 6.1 and 6.2). Furthermore, it is possible that this broad 20-year feature may simply reflect some aspect of the timescales of atmospheric–ocean inter-actions, or more likely a variable combination of all three phenomena (see Section 7.3).

As far as longer periodicities are concerned, proxy data help to build up a rather stronger case for the 80- to 90-year and 180- to 200-year cy-cles. But again, these periodicities come and go in an erratic manner. More important, they are the only source of information for periodicities longer than a few hundred years. In the millennial range the emergence of the 1450-year cycle is an intriguing new development, but it is too soon to say just how significant this will be in providing new understanding of the cli-mate. Above all, proxy data have provided strong support for the existence of longer-term periodicities associated with the variations in the Earth's orbital parameters.

These conclusions are set out in more detail in Table 4.1. Together with the results in Table 3.2, they provide the starting point for identifying the possible physical explanations for regular fluctuations in the climate. The first step in this process is to examine certain aspects of the current knowledge about the causes of climate change.

Chapter 5

The global climate

... what we are concerned with here is the fundamental
interconnectedness of all things.

Douglas Adams (*Dirk Gently's Holistic Detective Agency*)

The wide range of data that can be exploited in the search for weather cycles
must now be put in context. This database provides a considerable amount
of information about the scale and time-span of a variety of meteorological
variables. But to understand what insight, if any, these observed fluctuations provide about the overall cyclic behaviour of the weather, we need to
look more closely at how the global climate functions. For only when the
variability of the weather is analysed in terms of the processes that govern
the global and regional energy balance of the Earth's climate is it possible
to make a sensible assessment of the evidence of cycles. In particular, the
possibility that all the ups and downs in the weather are nothing more than
the natural variability of the complex non-linear connections between various components of the climate must be explored in detail. Once this issue
has been examined it will be possible to focus on the more precise question
of cyclic behaviour.

This approach cannot go over all the standard climatological ground
covered by text books (see the bibliography). Instead, it will concentrate on
those features of the global climate that are apparently most closely linked
with the fluctuations identified in Chapters 3 and 4. This approach means
focusing on the slowly varying components of the global weather machine
that are the most obvious factors in fluctuations from year to year. These
are the components that can alter the energy balance in different parts of
the world for long periods. They include snow cover, polar pack ice and
sea surface temperatures (SSTs). In addition, this analysis must consider

how these various elements of the climate both influence the weather and interact with one another. Throughout, the central observation will be that everything is connected to everything else, often in the most complicated way. So while it may help to identify the most obvious connections, any assumptions about these behaving in a predictable and simple manner must be treated with great caution.

5.1   Circulation patterns

The key to identifying the origin of possible weather cycles is to find out what causes the large-scale global atmospheric circulation to shift from year to year. As has emerged in the last two chapters, the most convincing examples of cyclic or quasi-cyclic behaviour involve changes over large areas. Such extensive and long-lasting shifts are linked to the circulation patterns that become established each year. To understand how day-to-day weather systems are an integral part of these patterns we need to look at the middle and upper levels of the atmosphere. The reason for doing this is that it is possible to strip away the complication of ground-level weather features and reduce the associated effects of land masses and reveal the bare bones of global circulation.

The standard approach is to show the average height of a given pressure surface (e.g. 500-hPa surface) rather than using surface pressure maps. This contour map (see Fig. 5.1a) shows that in the northern hemisphere in winter the normal circulation is dominated by an extensive asymmetric cyclonic circumpolar vortex with the primary centre over the eastern Canadian Arctic and a secondary one over eastern Siberia.[1] In summer (see Fig. 5.1b), the pattern is similar, but the vortex is much less pronounced. The important feature of this circulation as far as we are concerned is that major troughs and ridges form in this vortex. At any time the number of troughs and ridges may vary, as will their position and amplitude. These are known as 'long waves' (or 'Rossby waves' – after the Swedish meteorologist Carl Gustav Rossby, who first provided a physical explanation of their origin).

The basic physical explanation of the existence of long waves in the upper atmosphere is linked to the circulation of transient eddies at lower levels. The combined spin of one of these low pressure systems and its

[1] The climatological data presented here use reanalysis data from either the ECMWF project (see http://www.ecmwf.int/data/era.html) or the NCEP/NCAR project (see Kalnay *et al.*, 1996). For a more general discussion of how these patterns underlie many features of global climatology see the general texts cited in the bibliography.

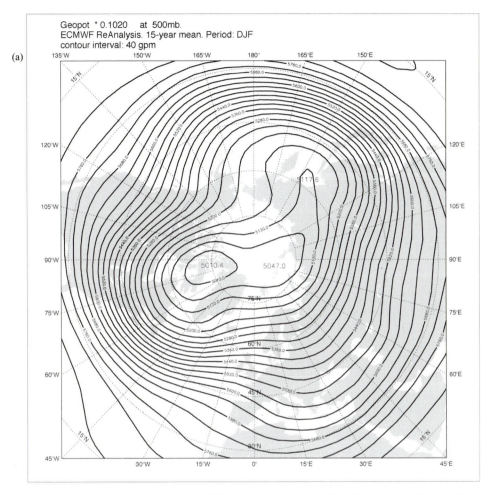

Fig. 5.1. The mean height of the 500-hPa pressure surface (in decametres) in (a) winter and (b) summer for the northern hemisphere. (With permission of the ECMWF.)

rotation about the Earth's axis (known as *vorticity*) effectively remains constant. So, a system in a stream of air moving towards polar regions will tend to lose vorticity as the distance from the Earth's axis of rotation reduces. This is compensated by the flow adopting an anticyclonic curvature (clockwise in the northern hemisphere) relative to the surface at higher latitude. Eventually, the curved path takes the airstream back towards the equator. As it reaches lower latitudes the reverse effect will take place as the distance from the axis of rotation increases and the airflow will adopt a cyclonic curvature. So the airstream tends to swing back and forth as it circulates the Earth. The two major troughs in the climatological mean flow are around 70° W and 150° W. Their position is linked to the impact of the major mountain

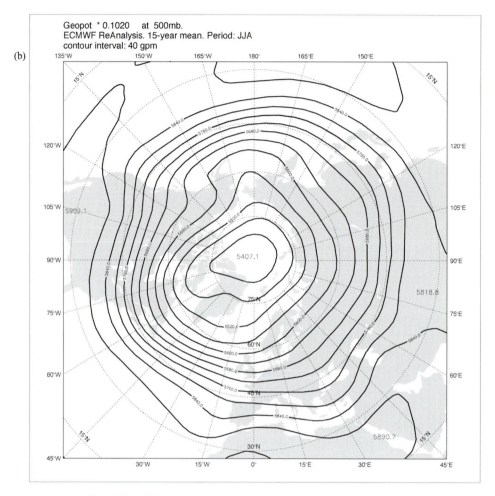

Fig. 5.1. (*cont.*)

ranges on the flow, notably the Tibetan plateau and the Rocky Mountains, together with the heat sources such as ocean currents (in winter) and land masses (in summer). Because the southern hemisphere is largely covered by water, the pattern is much more symmetrical and shows less variation from winter to summer (see Fig. 5.2a and b).

Understanding these patterns is central to analysing long-term weather fluctuations. What is not evident from these broad patterns is the nature of winds in the upper atmosphere. The contour maps record the thickness of the atmosphere between sea level and the 500-hPa surface, and this thickness is proportional to the mean temperature – low thickness values correspond to cold air and high thickness values to warm air. This means that the circumpolar vortices (see Figs. 5.1 and 5.2) reflect a poleward decrease in temperature and this drives strong westerly winds in the upper

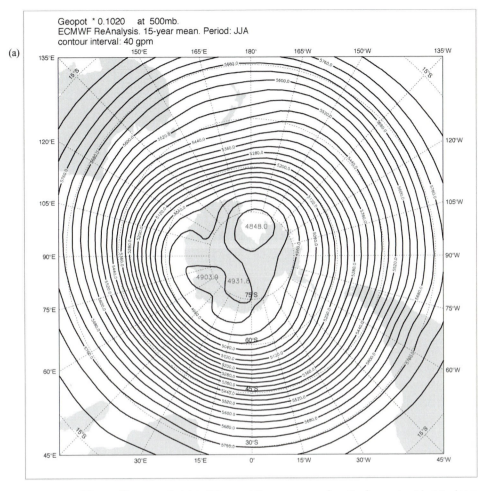

Fig. 5.2. The mean height of the 500-hPa pressure surface (in decametres) in (a) winter and (b) summer for the southern hemisphere (with permission of the ECMWF).

atmosphere. But, while the thickness decreases gradually with increasing latitude, the strongest winds are concentrated in a narrow region, often situated around 30° of latitude at an altitude between 9 and 14 km. The concentrated wind flow is known as the 'jet stream', and reaches maximum speeds of 160 to 240 kph, and can exceed 450 kph in winter. The reason for the restriction of the winds to a narrow core is not fully understood. Moreover, the structure can be complicated, especially in the northern hemisphere winter when the jet stream often has two branches (the subtropical and polar front jet streams). The mainstream core is associated with the principal troughs of the Rossby long waves. This circulation governs the movement of surface weather systems, so it is an important factor in understanding day-to-day weather patterns and longer-term fluctuations.

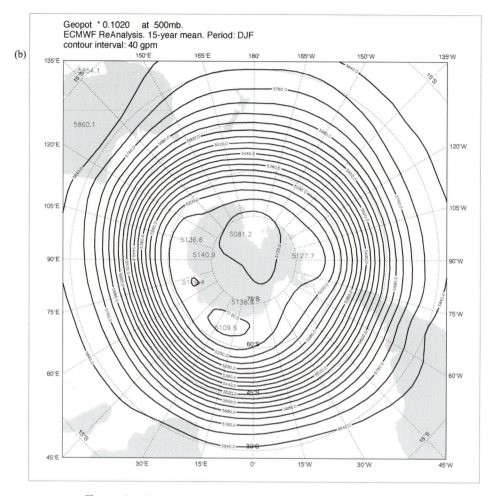

Fig. 5.2. (*cont.*)

Circulation patterns, which are radically different from those in Figs. 5.1 and 5.2, can last a month or two. They occur irregularly but are more pronounced in the winter when the circulation is strongest. Ridges and troughs can become accentuated, adopt different positions and even split up into cellular patterns. An extreme example of such a pattern occurred in the winter of 1962/63. Figure 5.3 shows the mean height of the 500-hPa surface in January 1963. The striking features are a pronounced ridge off the west coast of the United States (40° N, 125° W) and a well-defined anticyclonic cell just to the south of Iceland (60° N, 15° W). This set the stage for the extreme weather. Cold arctic air was drawn down into the central United States and into Europe. Conversely, warm tropical air was drawn far north to Alaska and western Greenland. The net effect was that

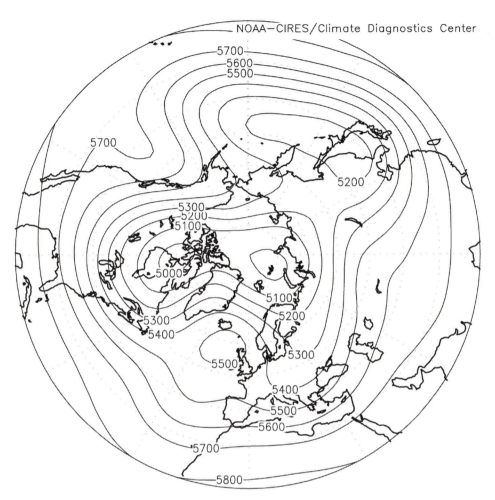

Fig. 5.3. The mean height of the 500-hPa pressure surface (in decametres) for January 1963, showing the pronounced wave pattern in mid latitudes due to 'blocking' off the west coast of the United States and close to the British Isles. (Data from NOAA Climate Diagnostic Center.)

while an extreme negative anomaly of $-10\,^\circ$C for the month was observed in Poland, an equal positive departure occurred over western Greenland.

Such an extreme pattern, which is usually termed 'blocking', shows how important it is to understand the causes of anomalous atmospheric behaviour.[2] Because the spacing incidence of extreme seasons, like the winter of 1962/63, can exert a major influence on the evidence of periodicities in weather series (see Appendix A.2), identifying their origin is a central requirement to providing an explanation of both climatic variability and

[2] See Rex (1950), Lejenas & Okland (1983) and Lejenas (1995).

possible weather cycles. So we need to address the basic questions of what causes the number, amplitude and position of the Rossby waves to change and then to remain stuck in a given pattern for weeks or months.

Clearly, in the northern hemisphere the distribution of land masses and the major mountain ranges play a dominant role. But this does not explain why the number of waves around the globe may range from three to six or why they can vary from only small ripples on a strong circumpolar vortex to exaggerated meanderings with isolated cells. It follows from the description of the requirement to conserve vorticity in the high-level flow that an important factor is the speed of the upper atmosphere westerlies. There is some evidence that when the wind speeds assume a critical value this enhances the chances of strong standing waves building up downstream from the troughs at 70° W and 150° E. But this begs the question of what governs the changing speed of the winds from year to year and within seasons.

Any explanation of what causes different circulation patterns must also show why their incidence changes. In winter in the northern hemisphere just four or five categories of circulation regimes occur about three-quarters of the time.[3] This fact may hold the key to understanding what controls the switches between regimes, as the limited number of states suggests there are tight constraints on which patterns are permissible in strong circulation systems. Any insight may then be extended to other times of the year and other parts of the world where circulation regimes exercise less influence on the climate. So, the essential issues are why a given regime becomes established, what causes it to shift suddenly to a different form, and above all what defines the incidence of given regimes.

To see what factors could play a part in setting up these patterns we must turn to the slowly varying components of the climate system. In doing so, however, it is essential to keep in mind one physical feature about the sources of energy that drive the weather. This is that the global climate can be regarded as a heat engine whose source of energy is solar radiation and whose sink is the energy radiated to space by the atmosphere and the surface of the Earth. Because most of the energy input is at low latitudes while a significant proportion of the outgoing radiation is emitted at high latitudes, there is a net transport of energy away from equatorial regions. More immediately relevant in terms of the upper atmosphere winds, the vast majority of the solar energy is absorbed at the surface and in the lower levels of the atmosphere. This means that the broad circulation patterns are driven by the energy from the bottom of the atmosphere, especially at low latitudes. So, although the upper atmosphere winds appear to be steering

[3] Palmer (1993).

the movement of the surface weather, they are essentially the product of the amount and distribution of the solar energy absorbed into the atmospheric heat engine.

The importance of this basic thermodynamic feature of the global climate cannot be underestimated. It does not rule out the possibility that changes in the upper atmosphere could exert a subtle influence on surface weather: so, in general, the tail does not wag the dog. Hence, we must concentrate first on those aspects of the climate that could result in the major flows into and out of the global atmosphere, which in principle are likely to lead to the most significant fluctuations in the weather. If these do not prove adequate explanations of the observed changes, then it will be necessary to turn to more complicated mechanisms. There is one other feature of the whole question of blocking that needs to be addressed before we embark upon mechanistic discussions. This is that, despite the fact many of the proposals we will consider are the product of computer modeling studies, the ability of the General Circulation Models (GCMs) used in climate change studies to reproduce the properties of blocking is not impressive. The conclusion reached in the Third IPCC Assessment Report[4] was that, while some models had managed to capture the location and seasonality of blocking in the northern hemisphere, the frequency and duration of the phenomenon was underestimated. Furthermore, the coupled ocean–atmosphere GCMs that are essential to studying the slowly varying components of the climate have not been tested in this respect.

5.2 Radiation balance

It follows from the observations about atmospheric circulation patterns that the radiative balance of the atmosphere and the Earth's surface, both globally and locally, play a fundamental part in controlling the weather. Because the physical processes involved are central to so many features of the climate, it is essential to know how it varies over time, as it may be the key to many of the fluctuations considered in this book. At the simplest level, over time the amount of solar radiation absorbed by the Earth must be balanced by outgoing heat radiation to space. But the balance involves a host of different effects.

The proportion of incoming solar radiation absorbed by the Earth depends on the absorption, reflection and scattering properties of the atmosphere and the surface. While a large number of observations have been made of these properties, they tend to provide only a piecemeal picture

---

[4] IPCC (2001), see Chapter 8, p. 506.

of their overall impact on the climate. Satellite measurements have, however, started to produce accurate pictures.[5] These have given values of the amount of solar radiation reflected or scattered into space without change of wavelength (the albedo) and also the amount of heat radiated to space as a function of season and of latitude and longitude. In particular, these results have started to unravel the puzzle of the role played by clouds in regulating the radiative heating of the planet.

Where there are no clouds the oceans are the darkest regions of the globe. They have albedo values ranging from 6% to 10% in the low latitudes and 15% to 20% near the poles. Ocean albedo increases at high latitude because at low Sun angles water reflects sunlight more effectively. The brightest parts of the globe are the snow-covered Arctic and Antarctic, which can reflect over 80% of the incident sunlight. The next brightest areas are the major deserts. The Sahara and the Saudi Arabian desert reflect as much as 40% of the incident solar radiation. The other major deserts (the Gobi and the Gibson) reflect about 25% to 30%. By comparison, the tropical rain forests of South America and Central Africa, as the darkest land surfaces, have albedos from 10% to 15%.

The pattern of outgoing long-wave radiation is more systematic. This reflects the fact that the temperature of the surface and the atmosphere decreases relatively uniformly from the equator to the poles. In effect these surfaces are close to being black bodies when it comes to emitting long-wave radiation and the emissivity of various surfaces does not vary appreciably. The average amount of energy radiated to space decreases from a maximum of 330 W m^{-2} in the tropics to about 150 W m^{-2} in the polar regions.

Before reviewing the impact of clouds on the energy balance both globally and regionally, it is necessary to consider the implications of the clear sky albedo measurements for long-term fluctuations in the weather. The most highly reflecting areas are also among the most variable climatic elements. For instance, increases in the extent of winter snow and ice cover in polar regions will have a significant effect on the amount of sunlight reflected into space. Similarly, but somewhat surprisingly, an expansion of the major deserts of the world will lead to an increase in albedo and hence to more solar radiation being reflected into space. So while deserts are regarded as hot places their expansion could lead to a general cooling, unless associated with some compensating changes in cloudiness.

The global distribution of clouds shows that they are most common over the mid-latitude storm tracks of both hemispheres. They also have a

[5] See for example Ramanathan *et al.* (1989) and Rossow & Schiffer (1999).

less striking maximum over the tropics, especially the region around south-
east Asia. On average, about 65% of the Earth is covered by clouds at any
given time. Clouds are almost always more reflective than the ocean surface and the land except where there is snow. So where clouds are present
they reflect more solar energy into space than do areas that have clear skies.
Overall, their effect is approximately to double the albedo of the planet from
what it would be in the absence of clouds to a value of about 30%. Conversely, where clouds are present less thermal energy is radiated to space
than where the skies are clear. It is the net difference between these two
effects that establishes whether the presence of clouds cools or heats the
planet. The scale of the blanket-like warming effect depends on the thickness of the clouds and the temperature of their tops. High clouds radiate
less than low clouds, and thick clouds are more efficient radiators than thin
clouds.

Satellite observations[6] from the Earth Radiation Budget Experiment
(ERBE) concluded that the net impact of clouds globally is to reduce the
amount of solar radiation absorbed by the Earth by 48 W m^{-2} and to reduce
the outgoing heat radiation to space by 31 W m^{-2}. Subsequently a consensus has settled on figures of -50 W m^{-2} for the total short-wave cloud forcing and 30 W m^{-2} for the long-wave cloud forcing,[7] giving a net figure of
-20 W m^{-2} for the net global cooling from a mean global cloud cover of
62%. It is, however, a measure of the uncertainties in this area that recent
work, using improved calibration and cloud detection sensitivities and the
International Satellite Cloud Climatology Program (ISCCP) data set, has
produced a figure of 67.5% for global mean cloud cover.[8] Furthermore, the
Third IPCC Assessment Report[9] shows that a review of predictions of the
models used to predict global climate change produce equivocal results as
to whether, in a warmer world, changes in cloud cover will have a warming or cooling effect. So, while overall clouds have a cooling effect on the
climate, their role in current climate change and weather cycles is still a
matter of considerable doubt.

When we turn to the role of clouds in regional climatology and in feedback mechanisms associated with climate change the position becomes
even more difficult to resolve. The blanketing effect of clouds reaches peak
values over tropical regions and decreases towards the poles. This is principally because clouds rise to a greater height in the tropics, and the cold
tops of deep clouds radiate far less energy than shallower, warmer clouds.
So where there are extensive decks of high cirrus clouds, the amount of

[6] Ramanathan *et al.* (1989).   [7] Kiehl & Trenberth (1997).
[8] Rossow & Schiffer (1999).   [9] IPCC (2001), see Chapter 7, Figure 7.2.

energy radiated to space, compared with clear skies in the same regions, is reduced by 50 to 100 W m^{-2}. These thick high clouds occur in three main regions. The first is tropical Pacific and Indian Oceans around Indonesia and in the Pacific north of the equator, where rising air forms a zone of towering cumulus clouds. The second is the monsoon region of Central Africa and the region of deep convective activity over the northern third of South America. The third is the mid-latitude storm tracks of the North Pacific and North Atlantic Oceans.

The pattern of increased albedo due to clouds is different. The regions associated with the tropical monsoon and deep convective activity reflect large amounts of solar radiation, often exceeding 100 W m^{-2}, as do the clouds associated with the mid-latitude storm tracks in both hemispheres and the extensive stratus decks over the colder oceans. The important difference is that these clouds at high latitudes have less impact on the outgoing thermal radiation as the underlying surface is colder and hence emits less energy whether or not there are clouds. So in the tropics the net effect of clouds is effectively balanced out, but over the mid- and high-latitude oceans polewards of 30° in both hemispheres, clouds have a cooling effect. This negative effect is particularly large over the North Pacific and North Atlantic, where it can be between 50 and 100 W m^{-2}.

The whole issue of trends in global and regional cloudiness remains the subject of scientific argument. Broadly speaking, surface measurements throughout the twentieth century show increases in cloud cover of a few per cent over most continents. These rises are negatively correlated with the corresponding decline in the diurnal range of surface air temperature that has been a feature of global warming over the last 100 years or so. Over the oceans, the general trend in recent decades appears to show a rise in cloud cover,[10] with a long-term upward trend in altostratus and nimbostratus across the mid-latitude North Pacific and North Atlantic Oceans and variations in the frequency of low cloud types across the Pacific and Indian Oceans related to El Niño Southern Oscillation (ENSO) (see Section 5.4).

Although weather satellites have been collecting regular pictures of clouds since the early 1960s, which might be expected to provide a continuous record of global cloudiness, this is not the case. Problems arise from the optical properties of different types of clouds and the changing sensitivity of satellite equipment during its lifetime. As a result, most of the work on using satellite estimates of changes and variations in cloud amount and type is not considered sufficiently reliable to be used in climate change studies. The ISCCP data set has, however, produced some useful results.[11] These show,

---

[10] Henderson-Sellers (1992).   [11] IPCC (2001), see Chapter 7, figure 7.2.

over the period from 1983 to 1994, a globally increasing trend in monthly mean cloudiness reversed during the late 1980s and early 1990s. There now appears to be an overall trend toward reduced total cloud amounts over both land and ocean during this period.

While these results provide a hint of the potential implications of fluctuations in global cloudiness for both climate change and weather cycles, they fall well short of evidence of real effects. The main difficulty now is to identify the reasons for the observed changes in the ISCCP data. In particular, the impact of the ENSO warm events in 1983, 1987 and the early 1990s appear to have exerted a significant influence on cloud amounts. Furthermore, the eruptions of the volcanoes El Chicon in 1982 and Mount Pinatubo in 1991 may also have played a part in shorter-term variations, especially for high-level cirrus. So, when it comes to measuring cloudiness, it is truly a case of everything being connected to everything else. This will become even more evident when we come to look at the possibility of solar activity affecting global cloudiness (see Section 6.3).

5.3 Prolonged abnormal weather patterns

The possibility of long-term changes in global cloudiness leads naturally into the consequences of abnormal weather patterns. This in turn links into many aspects of the attempts to conclude whether periodic behaviour in the weather is tied into the underlying tendency of the climate to have a 'memory' (see Section 2.7). Because a sustained period of extreme weather can produce more lasting changes in various components of the global climate, it follows that the scale and significance of these changes need to be assessed. As noted earlier, the most obvious factors are winter snow cover and the extent of polar pack ice: the even more important issue of SSTs will be considered separately in the next section. Here, we will concentrate on the most basic issue of whether a prolonged spell of extreme weather can affect the underlying components of the climate long enough to influence the weather in subsequent seasons, and conceivably lead to periodic fluctuations.

The possibility of extreme snow cover prolonging winters in parts of the northern hemisphere has been studied in terms of severe winters.[12] This analysis has been extended to cover fluctuations in the overall extent of the annual average snow cover in the northern hemisphere having longer-term consequences. The reason for this interest is obvious. Because snow is an efficient reflector of sunlight it reduces the amount of solar energy

[12] Namias (1985).

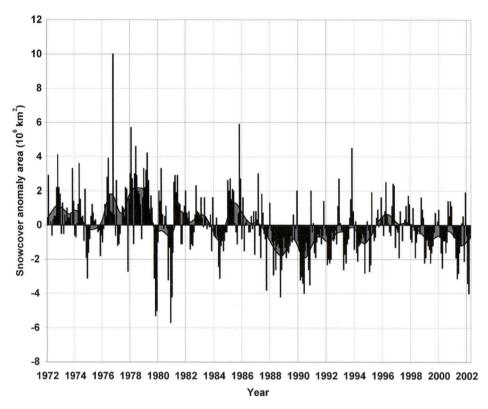

Fig. 5.4. Changes in snow cover in the northern hemisphere between January 1972 and May 2002 (vertical bars and heavy line). These fluctuate dramatically from month to month and show a long-term decline in snow cover that is closely correlated with the rise in temperature in the northern hemisphere north of 30° N. (Data from NOAA.)

absorbed at the Earth's surface. Hence, more extensive snow cover has a cooling effect. In principle, this cooling effect should lead to a colder climate and so produce more snow. This could lead to a positive feedback mechanism that drives the climate into a much colder regime. Conversely, a series of mild winters, which result in well-below-average snow cover, could reinforce a warming trend.

To have a lasting climatic impact, changes in snow cover in the northern hemisphere must last an appreciable time. The area covered by snow varies by a factor of over ten between winter and summer. So, to have a significant effect, anomalies must last well into the spring and summer. Satellite records show, however, that from month to month abnormal snow cover is a relatively transient phenomenon (Fig. 5.4).[13] These observations

[13] Details of northern hemisphere snow cover can be found on the NOAA website at: http://www.cpc.ncep.noaa.gov/products/monitoring_and_data/us_snow.html.

suggest that fluctuations in snow cover are not sufficient to lead to periodic behaviour. But in the longer-term, if they are associated with changes in the amount of sunlight falling at high latitudes during the summer, they become a major climatic factor (see Section 6.4).

At a regional level, changes in snow cover may exert a greater influence. It has been proposed that the extensive snow cover across Europe during the exceptional winter of 1962/63 (see Section 5.1) could have been partially instrumental in sustaining the cold weather. This winter was the coldest since 1830 and resulted in virtually all of Europe north of the Alps being covered in deep snow throughout January and February 1963. The prolonged snow cover helped sustain the high pressure region over Scandinavia, which was the principal feature of the abnormal weather patterns. Similarly, in the United States the extensive snow cover that built up during the record-breaking cold spell in December 1983 helped prolong the wintry weather. During January 1984 it is estimated that in parts of the Great Plains the daytime maximum temperature was 5 °C lower than would have been expected on the basis of the prevailing atmospheric conditions.[14]

Although abnormal snow cover may sustain a cold winter for several weeks longer than might normally be expected, this does not appear sufficient to affect conditions for months and years ahead. Recent studies suggest, however, that autumn snow cover over Siberia may have a more lasting influence.[15] When it is more extensive than usual there is a significantly greater chance of the following winter being colder than normal in Europe and eastern North America. Conversely, when the cover is below normal the winter is likely to be mild in these areas. In effect, the Siberian autumn snow cover appears to have a significant influence on the winter North Atlantic Oscillation (NAO).

This observation may explain why, although in any particular part of the world there is little evidence of reliable rules about sequences of abnormal seasons, there is some evidence of more predictable behaviour within seasons. For example, in the British Isles the chances of weather in January persisting through February into March significantly exceed what would be expected on the basis of chance.[16] The same phenomenon occurs from July through to September. But for the rest of the year there is little evidence of persistence, with a complete breakdown in late spring and autumn. So all that can be said is that in the British Isles once winters or summers settle into a given pattern there is a relatively higher probability that it will get stuck in a rut until global circulation patterns alter with the progression of the seasons.

[14] Namias (1985).   [15] Saito & Cohen (2003).   [16] Gordon (1976) and Dyer (1978).

The progress of seasonal forecasts provides further insight into understanding fluctuations from season to season and year to year. Up until the late 1980s, the performance showed at best only marginal improvements in skill over chance in forecasting temperature. Moreover, results varied for different times of the year and for different parts of the northern hemisphere and different types of forecasts. Anticipating extremely cold or mild winters was the best bet over North America, especially for the eastern half of the continent. At other times of the year the forecasts were, in general, less successful, while predictions of near-normal temperatures were particularly unsuccessful.

In the 1990s, this work was transformed with the recognition that ENSO events could exert a major influence on seasonal weather patterns in the tropics, and possibly at higher latitudes. In terms of the United States, however, while winter forecasts have made progress, summer forecasts have not been that impressive. In part, this was a result of what became known as the 'spring predictability barrier' that afflicted ENSO forecasts (see Section 5.6). Recent work has shown that a remarkable increase in prediction skills can be achieved by optimal use of information about the SST anomalies in all the major ocean basins.[17] What this showed was that different ocean basins made different contributions to different parts of the country in each season. The results of this progress in seasonal forecasting work do, however, provide an insight into the problem of weather cycles. This is that the predictability of weather patterns appears to vary with the seasons and from place to place. This suggests that the likelihood of establishing longer-term patterns, which could lead to oscillatory behaviour in the weather, may also exhibit temporal and spatial variability. Indeed, the evidence presented in Chapters 3 and 4 shows a marked tendency to such behaviour. So although these observations on the possible impact of abnormal snow cover and on seasonal forecasts relate to only relatively short-term effects, they are a useful pointer to the problems of postulating the causes of longer-term climatic fluctuations. They also form a natural bridge to the new 'holy grail' for seasonal forecasting: the influence of ENSO and other oscillations on the weather around the world (see Section 5.4 *et seq.*).

Before moving on to ENSO, however, there is one further aspect of snow and ice to consider. This is the consequence of changes in the extent of polar pack ice. In some sectors of the northern hemisphere, notably the North Atlantic, the changes in ice cover can affect weather patterns. But, in general, the scale of these is small compared with the variations in snow cover (Fig. 5.4 and 5.5a). On average, the area of arctic pack ice ranges from a minimum of around seven million square kilometres in late

[17] Lau, Kim & Shen (2002).

(a)

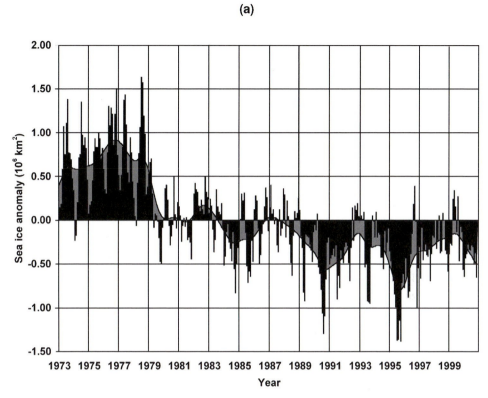

Fig. 5.5. Sea ice extent anomalies relative to 1973–2000 for (a) the northern hemisphere and (b) the southern hemisphere. (Data from UKMO.)

summer to a maximum of some 15 million square kilometres in early spring. The scale of these changes is, however, small compared with the variations in snow cover, which reach a maximum extent of around 45 million square kilometres in late winter and decline to less than 5 million square kilometres in late summer.

In the southern hemisphere the reverse is true. Because the Antarctic snow cover is permanent, and winter snow in South America, Australia and New Zealand is of limited extent, the most important variations are associated with the extent of Antarctic pack ice. The annual cycle has an amplitude of some 15 million square kilometres from a maximum extent of about 18 million square kilometres and around 3 million square kilometres in late summer. From year to year the extent of the ice cover can fluctuate by several million square kilometres.

Satellite measurements of the extent of both Arctic and Antarctic ice have been available since 1973 (see Figs. 5.5a and 5.5b). Arctic data over the past two decades show a decrease of 2.9% ± 0.4% per decade in sea ice extent. This decrease is strongest in the eastern hemisphere and has

(b)

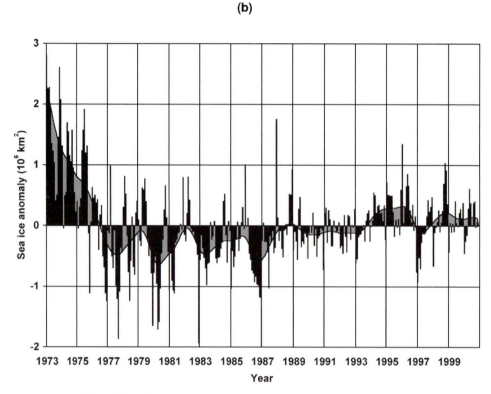

Fig. 5.5. (*cont.*)

been most apparent in summer. Sea ice extent in the Antarctic shows a weak increase of 1.3% ± 0.2% per decade since the late 1970s. Prior to this there was a marked reduction in the early 1970s.[18] Records are not sufficient to construct a longer time series for the Antarctic, but whaling ship logs suggest significantly greater ice extent in the Southern Ocean during the 1930s and 1940s than during recent decades, followed by a decline of some 25% in the extent of the ice (a reduction in area of 5.65 million km²) between the late 1950s and the early 1970s.[19] Possibly more intriguing is that since 1973 there has been a puzzling tendency for the anomalies to vary out of phase. The possibility that the extent of polar sea ice in each hemisphere could be linked in some way is also found in the analysis of millennial oscillations in the ice core data.[20] These results provide yet further evidence of the potential complexity of the interconnectedness of the global climate.

The climatic significance of snow and ice cover changes depends on how other components of the climate system will respond to them. For

[18] Cavalieri *et al.* (1997).   [19] de la Mare (1997).   [20] Steig & Alley (2002).

instance, the shift in the extent of Antarctic pack ice could produce parallel shifts in the storm tracks at lower latitudes and hence alter the cloud cover. Depending on the form of these additional responses, the net effect could be to either reinforce the changes in the pack ice extent or largely cancel them out (this is all a matter of whether the feedback is positive or negative – see Section 7.3). More important may be how these changes couple with shifts in the behaviour of the oceans, both at high latitudes and in the tropics. In this context, the recent discovery of an Antarctic Circumpolar Wave (see Section 5.7) may be a clue to these teleconnections.

So far the emphasis has been on changes at high latitudes. In terms of the 'blocking' (abnormal weather) patterns considered, this is reasonable. But this approach must not be allowed to overlook the fact that the global atmosphere is driven principally by the energy absorbed in the tropics. So in terms of the overall analysis of abnormal weather patterns, it is essential to consider the global picture. This is particularly important when considering SSTs. For while the role of SST anomalies at high latitudes have frequently been invoked to explain abnormal weather patterns in these regions, it has become increasingly evident since the 1980s that the tropics matter more. So now we must turn to the blossoming subject of 'oscillations' in the tropics and, first and foremost, to ENSO.

5.4 El Niño Southern Oscillation (ENSO)

When considering year-to-year variations in the Earth's climate, possibly the most important factor is how heat is taken up, stored, carted around the world and released by the oceans. This is a consequence of the much greater heat capacity of the oceans. Large-scale temperature anomalies can last far longer than the more fleeting changes in the atmosphere, or in snow cover and pack ice. Whereas the latter are measured in weeks and months, the former can last for years or much longer. So the oceans have the capacity to exhibit fluctuations on the same timescales as the periodicities examined in Chapters 3 and 4. As such they may be the key to many of the apparently regular fluctuations in the weather.

The changes in the oceans cannot, however, be considered in isolation. They are linked with the effects that have been discussed in Sections 5.2 and 5.3. Long-term changes in cloudiness may affect how much energy the oceans, especially in the tropics, absorb. Shifts in the extent of pack ice may influence how rapidly cold dense water descends into the depths in polar regions. Sustained changes in precipitation may have similar effects (see Section 7.3). Because these changes can take decades or centuries before influencing the temperature of upwelling cold water at lower latitudes, or even the opposite polar regions, they have the capacity to establish long-term fluctuations. But most important of all is that changes in atmospheric conditions can lead to shifts in the oceans' surface, which in turn can alter the weather patterns. These ocean–atmosphere feedback mechanisms have the potential to set up oscillatory behaviour and so produce periodicities or quasi-periodicities in the weather, and have become the subject of intense scientific interest in the last decade or two.

The most celebrated of these quasi-cyclic ocean–atmosphere interactions is El Niño and its association with the Southern Oscillation.[21] Section 3.8 discussed the evidence of cycles in the Southern Oscillation. The links between this atmospheric pattern and the large-scale fluctuations in SSTs across the tropical Pacific are the key to El Niño. For this reason, the overall phenomenon is generally known as El Niño Southern Oscillation (ENSO).

The name El Niño comes from the fact that a warm current flows southwards along the coasts of Ecuador and Peru in January, February and March: the current means an end to the local fishing season and its onset around Christmas means that it was traditionally associated with the Nativity (El Niño is Spanish for little boy and also refers to the Christ Child). In some years, the temperatures are exceptionally high and persist for longer, curtailing the subsequent normal cold upwelling seasons. Since the upwelling cold waters are rich in nutrients, their failure to appear is disastrous for both the local fishing industry and the seabird population, and the term El Niño became associated with these much dramatic interannual events.

The importance of ENSO was noted by scientists following the 1972/73 event, but it was not until the record-breaking event in 1982/83 that El Niño started to become part of popular culture. Much of this attention focused on the warm events that were traditionally associated with the term El Niño, but during the 1990s it became apparent that intervening cool episodes, which were an integral part of ENSO behaviour, were of comparable meteorological interest. These intervening cool episodes have come to be known as La Niña (Spanish for little girl) and have to be considered when describing the complete ENSO phenomenon.

In normal circumstances, seen from the SST anomalies across the Pacific, an ENSO warm event follows a rather well-defined pattern (Fig. 5.6).

---

[21] A huge amount has been written about ENSO. For standard texts the reader is advised to start with Philander (1983) and graduate on to the same author's more comprehensive and technical book Philander (1990). For more a recent review of current understanding of the phenomenon, Diaz & Markgraf (2000) is as good a place as any to start.

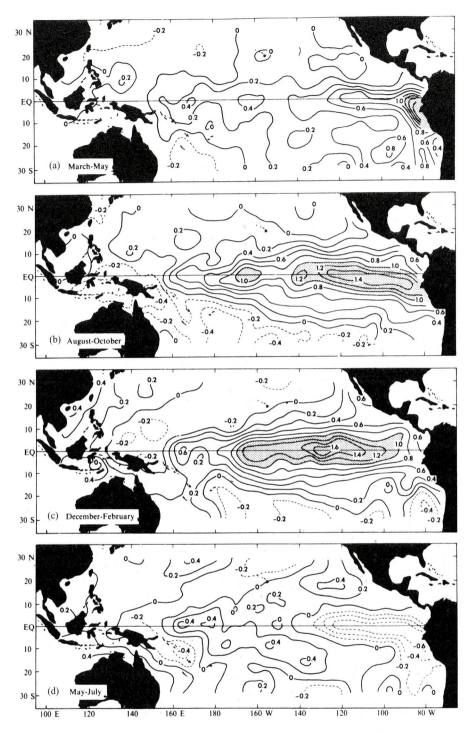

Fig. 5.6. Sea surface temperature anomalies (°C) during a typical ENSO event obtained by averaging the events between 1950 and 1973. The progression shows (a) March, April and May after the onset of the event; (b) the following August, September and October; (c) the following December, January and February; and (d) the declining phase of May, June and July more than a year after the onset. (From Philander, 1983. With permission of Macmillan Magazines Ltd.)

In the ocean above-average surface temperatures off the coast of South America mark the onset in March to May. This area of abnormally warm water then spreads westwards across the tropical Pacific. By late summer it covers a huge narrow tongue stretching from South America to New Guinea. By the end of the year the centre of the elongated region of warm water has receded to around 130° W on the equator and temperatures are returning to normal along the coast of South America. Six months later the warm waters has largely dissipated and in the eastern Pacific has fallen below the climatological normal.

In parallel with these changes in SSTs, large atmospheric shifts are in train. The surface pressure, wind and rainfall reveal that starting in the October and November before the onset of a warm event, the pressure over Darwin, Australia, increases and the tradewinds west of the dateline weaken. At the same time, the rainfall over Indonesia starts to decrease, but near the dateline it increases. In addition, the narrow band of rising air, cloudiness and high rainfall known as the Intertropical Convergence Zone (ITCZ), which girdles the globe, shifts position. Normally, it migrates between 10° N in August and September and 3° N in February and March. As a precursor to an ENSO warm event it shifts further south in the eastern Pacific, to be close to or even south of the equator during the early months of an El Niño year.

As the area of anomalous SST spreads westwards, a region of exceptionally high rainfall associated with the shift of the ITCZ accompanies it. During the mature phase of the event, most of the tropical Pacific is not only covered by unusually warm surface water but has also unusually weak trade winds associated with the southwards displacement of the ITCZ. Moreover, the heat transfer from the ocean means that the entire tropical troposphere in the region is exceptionally warm. This maintains the abnormal rainfall until the temperatures return to more normal values. With this return to normality the atmospheric patterns lapse back into a more standard form.

In parallel with these sea surface and atmospheric changes, important developments occur beneath the surface. The tropical Pacific can be regarded as a thin layer roughly 100 m thick, of warm light water sitting on top of a much deeper layer of colder denser water. The interface between these two layers is known as the 'thermocline'. High SSTs correlate with a deep thermocline and vice versa. As an ENSO warm event develops, the easterly trade winds that normally drive the currents in the equatorial Pacific become exceptionally weak. The sea level in the western Pacific falls and the depth of the thermocline is reduced (Fig. 5.7a). Intense eastwards currents between the equator and 10° N carry warm waters away from the

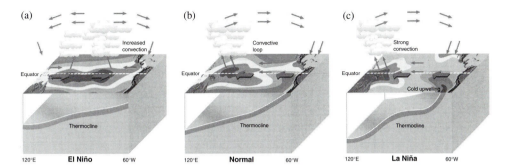

Fig. 5.7. Schematic illustration of the changes that take place between (a) El Niño conditions, (b) normal conditions and (c) La Niña conditions. The important features in (a) are the thermocline is less tilted than usual as the sea surface temperature is above normal in the eastern Pacific and there is increased convection in the central and eastern Pacific, whereas in (c) the reverse is the case as the thermocline steepens from west to east and convection moves farther west (with permission of WMO).

western Pacific. Along the western coast of the Americas there is an increase in sea level that propagates polewards in both hemispheres. This motion, which may be associated with cyclone pairs in the atmosphere north and south of the equator (see Section 5.8) that reinforces the early flow, propagates an eastward motion or wave. This called a 'Kelvin wave' (named after Lord Kelvin in recognition of his fundamental work in wave dynamics).

The changes in the ocean contain two important pieces of information. First, the observed movement in the oceans is a consequence of the alteration of the winds that normally drive the currents away from South America. This standard pattern produces lower sea levels in the east than in the west. It also means that cold water is drawn in from higher latitudes and also from greater depths. As the wind weakens so does the current. Sea levels rise and warm water spreads back to cover the cold water. This leads to a second counter-intuitive observation. The development of the SST anomaly appears to reflect a westward movement of warm water: that is not the case. The anomaly first appears off the coast of Peru, reflecting the fact that a small reduction in the overall movement westwards can lead to the cold Humboldt Current being capped by warmer waters. But the much more extensive region of abnormally warm water across most of the equatorial Pacific only develops after a sustained movement of warm water from the western Pacific. So, although the anomaly appears to move westwards, it is the counter-movement of water that is causing the observed effects. This underlines the central fact about all aspects of the ENSO that only by considering the combined atmosphere–ocean interactions is it possible to understand the overall behaviour of the phenomenon.

When the ENSO swings away from El Niño conditions it may either spend some time in what constitute normal conditions (see Fig. 5.7b) or pass rapidly on into to La Niña conditions (see Fig. 5.7c). At these times, the sea surface temperatures of the central and eastern equatorial Pacific Ocean are cooler than normal. Although less well-known than El Niño events, they are an essential part of the ENSO phenomenon. Just like El Niño, La Niña events disrupt the normal patterns of tropical convection and constitute the opposite phase of the Southern Oscillation circulation. The core feature of these events is that cooler than normal ocean temperatures develop in the central and eastern equatorial Pacific Ocean. Atmospheric pressure over the cooler water rises, but over Indonesia and northern Australia it falls below normal. Overall, the pressure difference across the Pacific increases and strengthens the easterly tradewinds.

While the nature of ENSO can be described in terms of changes in the tropical Pacific, its impact spreads far and wide.[22] The effects are most noticeable elsewhere in the tropics. During El Niño episodes, the normal patterns of tropical precipitation and atmospheric pressure are disrupted (Fig. 5.8). The distribution of rainfall shifts all around the globe as the changes associated with the Southern Oscillation alter the position of not only the ITCZ in the Pacific but also longitudinal atmospheric circulations. This leads to the region of heavy rainfall over Indonesia moving eastwards to the central Pacific. At the same time there is a smaller but significant move of the heaviest rainfall over the Amazon to west of the Andes. More important is that the region of ascending air over Africa is replaced by descending motion.

As the El Niño event develops from June through September, rainfall over parts of south-eastern Australia is often well below normal. Dryer than normal conditions are also observed over south-eastern Africa during the austral winter. The summer monsoon rains of India and China tend to be more erratic and often do not penetrate into north-west India and northern China. Over South America, unusual storm activity brings above normal rainfall to the central coast of Chile. Later, during the austral summer, frequent and often heavy rains deluge the subtropics east of the Andes, and the usually dry coastal regions of southern Ecuador and northern

---

[22] The early work on the global impact of ENSO was published by Ropelewski & Halpert (1987). More recent analysis of impacts is reviewed in Diaz, Hoerling & Eischeid (2001). In addition, updates of ENSO impacts can be found in the regular annual reviews published by the WMO.

(a) **Warm episode relationships June–August**

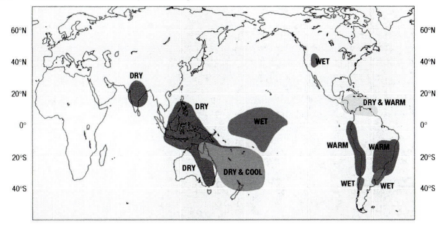

(b) **Warm episode relationships December–February**

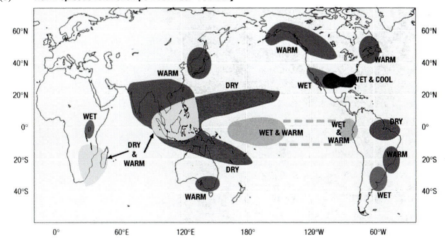

Fig. 5.8. Schematic diagrams of the areas that experience a consistent precipitation or temperature anomaly during an El Niño event in (a) the northern summer, and (b) the northern winter (with permission of WMO).

Peru often experience torrential rainfall. In contrast, much of the Amazon Basin and the north-eastern region of Brazil become drought stricken, and rainfall is reduced over Indonesia, Malaysia, the Philippines and northern Australia.

Another consequence of El Niño events is the impact on tropical storms around the world. The decline of hurricane activity in the tropical Atlantic and the Caribbean is well documented and has become an integral part of seasonal predictions of storms in this region. Over the north-western Pacific the number of tropical storms does not vary appreciably,

(a)   **Cold episode relationships June–August**

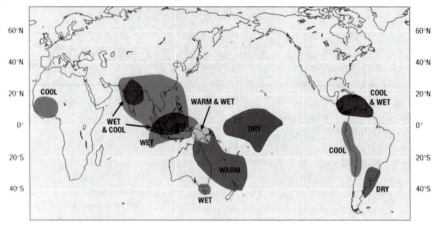

(b)   **Cold episode relationships December–February**

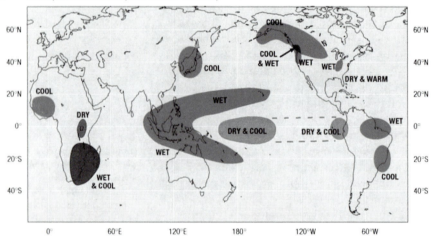

Fig. 5.9. Schematic diagrams of the areas that experience a consistent precipitation or temperature anomaly during a La Niña event in (a) the northern summer, and (b) the northern winter. (With permission of WMO.)

but the number of cyclones originating in the region west of 160° E declines, whereas the number forming from 160° E to just east of the dateline increases. This reduces the impact of such storms, including the total rainfall, on the Philippines, Japan and southern China. Also, tropical storm activity usually rises in the eastern Pacific and in the Southwest Pacific east of the dateline, while around Australia activity declines.

When the Pacific shifts into a La Niña mode there is a worldwide shift in weather patterns (Fig. 5.9). Cloudiness and rainfall over the central and eastern equatorial Pacific decline, especially in the northern hemisphere winter and spring seasons. At the same time, rainfall is enhanced over Indonesia, Malaysia and northern Australia, during the northern winter, and over the Philippines during the northern summer. Wetter than average conditions are also observed over south-eastern Africa and northern Brazil, during the northern winter season, while southern Brazil and central Argentina tend to be dryer than normal in their winter season. Also, during the summer, the Caribbean and northern South America are usually cooler and wetter than normal. Furthermore, there is a pronounced impact during the austral winter and spring over most of Australia, producing well above average rainfall, especially in the east of the continent. In the northern summer, the monsoon over India tends to be stronger than normal, especially in the north-west. Tropical storm activity in the tropical Atlantic tends to be the reverse of El Niño events with activity increasing when La Niña conditions prevail. In the Northwest Pacific the region of genesis shifts westwards with La Niña events.

Given the scale of these changes in the tropics, it is reasonable to assume that parallel extratropical disturbances occur. Somewhat surprisingly, a coherent set of connections between ENSO events and abnormal weather patterns at higher latitudes is less easy to find. One explanation for this is that a link between the anomalous circulation in the tropics and that of the mid latitudes requires the wavelength of these patterns effectively to be in tune. The mid-latitude Rossby waves (see Section 5.1) in the upper levels of the troposphere, which play a central role in defining the anomalous weather patterns of these latitudes, typically have a wavelength of about 6000 km and amplitude of some 3000 km. Because the wavelength and amplitude of these waves change with the seasons, becoming more pronounced in winter, it may be that the connections are most effective at this time of year, especially over the North Pacific and North America.

The proposed mechanism is that, during El Niño events, the increased heating of the tropical atmosphere over the central and eastern Pacific also affects atmospheric circulation patterns at higher latitudes. The jet streams shift their location and mid-latitude depressions are steered along different courses and tend to be more vigorous than normal in the eastern North Pacific. Abnormally warm, moist air is pumped into western Canada, Alaska and the extreme northern United States. Thunderstorms also tend to be more frequent in the northern Gulf of Mexico and along the south-eastern coast of the United States, resulting in wetter than normal conditions there. Conversely, during La Niña events, over the North Pacific mid-latitude low-pressure systems tend to be weaker than normal in the region of the Gulf of Alaska. This favours colder than normal air over Alaska and western

Canada, which often penetrates into the north-western United States. The south-eastern United States, by contrast, becomes warmer and drier than normal.

In practice, the correlation coefficients between the tropical Pacific and over North America are not high. Even over western Canada where the correlation is most pronounced, less than half the variance is attributable to events in the tropical Pacific. Since the strong El Niño event in 1972/73, some severe winters over North America, like 1976/77 and 1982/3, have coincided with warm events, and others, like 1977/78 and 1978/79, have not. During the 1990s the connections appeared to be becoming clearer. A successful winter forecast in 1991/92 based on foreseeing the coming El Niño events showed that these links could be used to good effect, and the spectacularly correct forecast of a very warm winter of 1997/98 over much of North America, linked to the exceptional El Niño event in 1997, was seen as confirmation of the strength of these connections. The fact that the next two winters were even milder, when a La Niña event was in full swing, took the gilt off the forecasting gingerbread (see also Section 5.6).

The differing response of the mid-latitudes to ENSO events raises the interesting question of just how important is the precise timing of the onset of, say, an El Niño warm event. In particular, how these changes link in with the biggest cycle of all – the annual cycle. For the major warming events since the 1970s, the time of onset and the rate of warming vary appreciably from event to event (Fig. 5.10). Whether or not the arrival of a warm event coincides with the normal annual warming of the eastern Pacific at around the turn of the year may be crucial to its wider impact. Because each event pumps so much energy into different parts of the tropical atmosphere, the timing of this huge impulse with respect to the march of the annual cycle can affect how much it influences seasonal weather patterns. A similar set of issues surround the timing of the onset of La Niña events.

The global impact of El Niño means that any evidence of cyclic behaviour will have important implications for weather cycles in general. The conclusions drawn from analyses of quasi-cycles in the timing of El Niño depend to a large extent on the period covered by the analysis. Where attention has focussed on the reliable monthly SST measurements available since around 1950 the results have concentrated on a combination of a quasi-biennial oscillation (QBO) and a broad periodicity around 4 to 5 years, sometimes referred to as a quasi-quadrennial oscillation (QQO).[23] Just how much of the variance is associated with these two periodicities

---

[23] Ropelewski, Halpert & Wang (1992) and Jiang, Neelin & Ghil (1995).

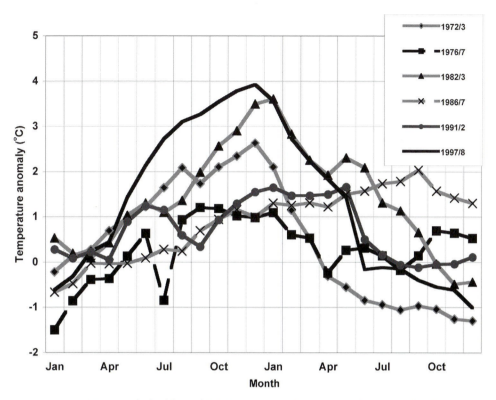

Fig. 5.10. The build-up of ENSO warm episodes since 1972 shows considerable variations from event to event. Nevertheless, they generally peak around the end of the calendar year, although in the case of the 1986/87 event it took two years to mature, while the others took about a year to reach their peak.

can be examined by filter analysis. If the observed SST anomalies for the 'NINO3' region (between 150° W and 90° W in the equatorial band 5° N and 5° S) are smoothed with two Morlet band pass filters centred on 24 and 48 months (see Appendix A.7) and these two signals used to reconstruct the original time series, a surprising amount of the observed variance can be attributed to the broad periodicities (Fig. 5.11). Furthermore, in spite of using filters with periods of precisely 2 and 4 years, whereas the spectral analysis of the NINO3 series shows the QBO and QQO to have periods of 26 and 53 months, the major peaks in the SST anomalies match within a month or two. This supports that the hypothesis that El Niño warm events are 'phase locked' to the annual cycle (see Sections 5.5 and 5.6) with all the significant events occurring between September and January, apart from the peak in June 1987. It also raises the issue of how much the QBO is phase locked with the annual cycle and how this is reflected in the quasi-cyclic behaviour of seasonal weather patterns.

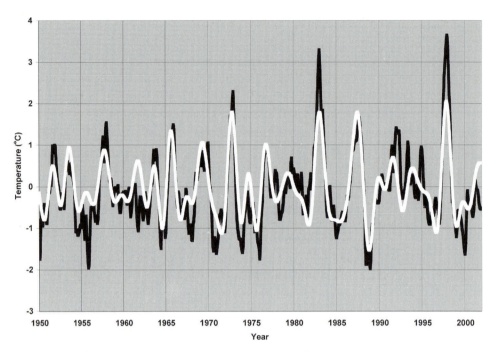

Fig. 5.11. A comparison of the observed bimonthly temperature anomalies in the NINO3 area (black line) and a reconstruction of this series using a combination of two filtered versions of the data (white line): one using a Morlet centred on 24 months and the other using a Morlet centred on 48 month. This result shows that the data can be largely represented by the combination of a quasi-biennial oscillation (QBO) and a quasi-quadriquennial oscillation (QQO) that is effectively 'phase locked' to the warm events peaking at around the end of the calendar year.

Wavelet analysis of SSTs in the tropical Pacific since the late nineteenth century (see also Section 3.8, and Fig. 3.8) presents a more complicated picture.[24] During the periods 1875–1910 and 1960–1990, the variance in the 2- to 8-year band was highly significant, but in the intermediate period it was largely absent, although there were sporadic outbreaks of what might be regarded as biennial behaviour in the 1920s and early 1930s. For the rest, the behaviour of ENSO, as reflected in these SSTs, lacks the sustained regularity to enable us to make any claims of periodicity. Instead, we see here, as elsewhere, spells of what look like periodic behaviour interspersed with more random episodes. Nevertheless, there is a widespread assumption that the rather ill-defined periodicity between 3 and 5 years found in many meteorological and proxy records is related to ENSO.

[24] Allan (2000).

5.5 Modelling El Niño and La Niña

The major ENSO events of the 1980s led to a wide range of models being developed to explain the observed changes. In terms of this book, perhaps the most illuminating approach to interdecadal fluctuations remains the one developed by Nicholas Graham and Warren White at the Scripps Institute of Oceanography at La Jolla, California, in the late 1980s.[25] This considered a natural oscillator of the Pacific Ocean and atmosphere system that produces irregular oscillations of around 3 to 5 years. The model proposed a set of relatively simple couplings between the tropical atmospheric circulation, the warm upper layer of the ocean and the SSTs in the eastern Pacific, which interact in a manner that has become known as the 'delayed oscillator' model. This approach was a refinement of earlier models that sought to address a basic problem. This is what causes the switch between El Niño and La Niña conditions, since both situations are maintained by winds that blow into an area of warm air rising over warm water. As has already been noted, in normal conditions the winds blow from the east towards the warmer water in the western Pacific. As these winds pile up warm water in the west they draw deeper colder water to the surface in the east, accentuating the temperature contrast that drives the wind that strengthens the contrast leading to a strong La Niña situation. So in theory these conditions should become a permanent feature of the Pacific.

During an El Niño the opposite feedback mechanism reinforces the anomalies. As unusually warm water extends eastwards, it is accompanied by winds from the west into the rising air over this warm water. As this cuts off the upwelling cold water it strengthens the westerly wind, enhancing the anomalous conditions. So, in principle, either El Niño or La Niña conditions could last indefinitely. But as the sea level rises in the east, it transmits a 'signal' to the west to lower the sea level in the west.

This 'signal' is in effect a travelling displacement of the thermocline. To understand how this signal can behave we need to consider what happens when the thermocline is disturbed. Hydrodynamical models show that in an ideal case of a symmetrical bell-shaped depression in the thermocline centred on the equator it will disperse into two waves – an eastward-travelling Kelvin wave and a westward-travelling Rossby wave. The former is a gravity-inertia wave that shows no meridional velocity variations and on which the

[25] Graham & White (1988). Other early influential work on models that has had considerable impact on developing thinking on ENSO behaviour include Zebiak & Cane (1987), Suarez & Schopf (1988) and Battisti & Hirst (1989).

restorative forces are due to the stratification of the ocean and the rotation of the Earth. The restorative force of the latitudinal variation of the Coriolis parameter, by contrast, governs the latter. As a consequence, its speed of propagation varies with the distance from the equator.

So changes in the depth of the thermocline, and by implication the thickness of the warm layer, can be separated into these two waves. This means the signal travelling westwards from the high sea level in the East Pacific is a Rossby wave – while El Niño is sustained by eastward-travelling Kelvin waves. The Rossby wave takes from several months to a few years to cross, the time taken depending on how far it is from the equator. In effect, it tends to cancel out the positive feedback that is sustaining the El Niño. Eventually, this process is sufficient to switch off the El Niño event and produces a return to more normal conditions. How these effects combine, however, to produce approximately oscillatory behaviour requires a more complicated combination of ocean–atmosphere interaction.

The important feature of the Graham and White model is how it expands on earlier work to include effects more distant from the equator. This approach combines developments restricted to within 2 to 3 degrees of the equator in what is often termed the 'equatorial waveguide' and affects up to 12° north and south of the equator. In the equatorial waveguide, processes are dominated by the zonal component of the wind stress that causes changes in the slope and thickness of the warm upper layer. The balance between these stresses and the weak Coriolis force results in downwelling and upwelling Kelvin waves moving rapidly eastwards. These waves are now monitored by an array of buoys across the tropical Pacific and orbiting satellites.[26] Measurements from these platforms produce a detailed picture not only of how positive and negative sea level anomalies propagate across the Pacific basin every 2 months or so, but also of how the depth of the thermocline varies at the same time.

Outside the equatorial waveguide the circulation of weather systems becomes more important in altering the thickness of the warm layer. With cyclonic wind fields the stress at the ocean's surface causes divergence with upwelling at the centre of the motion and a decrease in the thickness of the upper level. Conversely, anticyclonic wind fields result in convergence, downward vertical motion and a thickening of the upper layer. These processes of altering upper layer thickness are known as 'Eckmann pumping'. The consequence of these processes is to generate Rossby waves moving

[26] See Kessler *et al.* (1996) and McPhaden *et al.* (2001). Details of the latest developments of the TAO/TRITON buoy system and its latest results can be found on http://www.pmel.noaa.gov/tao/ and http://www.jamstec.go.jp/jamstec/TRITON/.

westward away from the regions of abnormal sea levels that depend on the latitude. These speeds are much slower than those of the equatorial Kelvin waves. Near the equator they cross the Pacific basin in about 9 months. Towards the poles, the time increases rapidly to be about 4 years at 12° north and south. In the past these waves were ignored. They now appear to be a key element in the explanation of events in the tropical Pacific.

The combination of these effects leads to the conclusion that the quasi-periodic appearance of warm water in the eastern Pacific (El Niño) is just one aspect of the system that operates as a natural coupled oscillator (the ENSO cycle) of the tropical Pacific Ocean and the atmosphere. The reaction of the surface wind field over the tropical Pacific to SST anomalies in the eastern and central regions produces two distinct oceanic responses. First, in the case of warm anomalies, abnormal westerly winds within the equatorial waveguide generate Kelvin waves that reinforce the SST anomaly in the eastern and central Pacific by increasing the thickness of the upper layer. Second, outside the waveguide, positive Eckmann pumping produces upwelling Rossby waves that propagate westward, reflect from the westerly rim of the Pacific basin into the waveguide and return as Kelvin waves. Eventually this process reverses the growth in the thickness of the upper layer and reverses the sign of the SST anomaly.

The reverse situation develops from abnormally low SSTs in the eastern and central equatorial regions of the Pacific. Abnormal strong easterly winds within the equatorial waveguide reinforce the existing SST anomaly and decrease the thickness of the upper layer. Outside the waveguide, negative Eckmann pumping produces downwelling Rossby waves that propagate westwards. These in turn are reflected in the same way as the upwelling waves associated with a warm anomaly and return as Kelvin waves. It follows that because these waves are associated with downwelling they reverse the thinning of the upper layer associated with the abnormal easterly winds and so eventually reverse the SST anomaly. This delayed negative feedback serves to switch off whatever anomaly that is initially in place and tends to set up an oscillation between El Niño and La Niña conditions.

A simple computer model of the Pacific basin has been able to produce a quasi-periodic oscillation in equatorial SSTs. These were irregular, occurring at intervals of about 3 to 5 years, and resembled observed fluctuations in the Pacific. This model has obvious limitations but provides important insights into how a simple linear model can produce a physically realistic mechanism that combines a coupled oscillator and delayed feedback. The fact that the model relies on linear interactions is important. As will be seen in Chapter 8, a non-linear approach would be intrinsically chaotic. So, although in the real world the physical connections would not be linear, this simple model does produce useful insights. Apart from its oscillatory behaviour, it suggests that when the oscillations are small the system has little predictability, being dominated by random fluctuations in the global climate. When the oscillations are large, the delayed feedback mechanism is the primary influence and the behaviour is more predictable. This may be the key to many transient examples of apparently cyclic behaviour in the weather. When a major disturbance develops, it can produce the right combination of conditions to set up delayed feedback. This may lead to the system going through a few 'cycles' before the effects die away. So in the longer term the cycle loses its predictability.

What this model appeared to show was that the ENSO cycle involves a long-lasting oceanic signal, which is 'in tune' with the dimensions of the region involved. The resultant quasi-cyclic behaviour could carry on for many years and keep reappearing at random when the conditions are right for a bout of activity. Subsequent work on the balance between the coherent cycle in the Pacific and the inherent noisiness of the climate system has suggested that the processes involved may involve a more subtle balance between signal and noise (see Section 7.7).

The importance of this model is that it provides apparently convincing insights into quasi-periodic ocean–atmosphere interactions. Although these may persist for a long time, they are restricted to part of the global climate. How the global climate system, including other ocean basins, could produce similar quasi-cyclic behaviour will require more detailed models. There is growing evidence of similar behaviour around the globe (see Section 5.7), which may link various ocean basins through wider atmospheric circulation patterns. So while these regional effects may provide persuasive evidence of 'delayed oscillator' mechanisms producing quasi-periodic behaviour, they are not yet the basis for more general predictions of weather cycles. Furthermore, any explanation of the natural variability of the climate must include the influence of extraterrestrial effects (see Chapter 6). But before we can examine these we need first to look at the capacity of the ENSO models to forecast developments in the Pacific months and years ahead, and then to bring in the evidence of oscillations in other parts of the world.

There is one other aspect of the non-linear interaction of ENSO behaviour that has become an integral part of modelling work. This is the influence of the seasonal cycle. The apparent phase-locking of the ENSO cycle to the annual cycle means that it may become organised around the mathematical structure known as the Devil's staircase.[27] This could

---

[27] Jiang, Neelin & Ghil (1995).

result in the frequency response of ENSO becoming locked into steps that are linked to the annual cycle according to a sequence of rational fractions. A consequence of this frequency locking is that the observed quasi-periodic behaviour may consist of a chaotic motion between the natural ENSO frequency and the best fit with the annual modulation, so, if this inherent frequency is normally between 4 and 5 years, it will hop from one to the other over time, or may even effectively miss a beat when the two cycles are completely out of phase. So, this motion can be irregular as the ENSO cycle is entrained non-linearly into synchrony with the annual cycle.

5.6 Forecasting ENSO behaviour

The first test of any model of ENSO behaviour is to produce forecasts of future development. The realisation during the 1980s of the fundamental importance of the ENSO in defining weather patterns throughout the tropics and possibly further afield led to a burst of activity in developing forecasting models using both empirical and physical techniques. Building on the physical insights obtained during the events of 1982/83 and 1997/98, modellers have been able to create a variety of detailed simulations of how the atmosphere and the ocean interact across the tropical Pacific. Because it is possible to exploit the relevant part of coupled ocean–atmosphere GCMs, which are used to study climate change, and empirical data of how past events behave, it has been possible to treat the tropical Pacific in isolation and to tune it response to changing conditions to build up a variety of modelling schemes. These hybrid systems have been used with varying degrees of success to provide regular forecasts of the occurrence and course of ENSO events.

The performance of these various models is the subject of continual review.[28] What has emerged so far is that the different models are all capable of providing useful forecasts. Consistent with the results of the general studies of the ENSO phenomena, the best results are obtained when strong episodes (both warm and cold) occur, but when the fluctuations are weaker

---

[28] A frank and up-to-date assessment of the current limitations of ENSO forecasting is given in Federov *et al.* (2003). To stay abreast of present-day developments it is probably best to use the various reports prepared by groups involved in ENSO forecasting (e.g. Bureau of Meteorology, Australia: http://www.bom.gov.au/climate/; ECMWF: http://www.ecmwf.int/services/seasonal/forecast/; and NCEP: http://www.cpc.ncep.noaa.gov/products/analysis_monitoring/index.html.)

or the equatorial Pacific is quiescent, the models were in trouble. Moreover, while all the models showed reasonable skill, their performance fluctuated with some doing better than others in certain circumstances, while at other times the reverse was the case. For instance, both the sustained moderate warm event between 1991 and 1995 and the cool event between 1998 and 2001 caught all the forecasters out. More disturbing was the fact that the physical models did not significantly outperform the simpler empirical models.

This slow progress reflects complex links between various cyclic and quasi-cyclic phenomena that are of particular relevance to this book. First, the whole question of the timing of the build-up of ENSO events and the annual cycle (see Sections 5.4 and 5.5) is a fundamental part of this work, which appears in the form of emerging events failing to develop as forecast in the early part of the year: a phenomenon that the forecasters have come to define as the 'spring predictability barrier'. This challenge is compounded by the existence of intraseasonal fluctuations in tropical weather, which although quasi-periodic in nature (see Sections 3.11 and 5.8), introduce a substantial and inherently chaotic element into the forecasting process.

When it comes to applying forecasts of future ENSO behaviour to predicting seasonal weather in other parts of the world, the possibility of quasi-cyclic behaviour in other parts of the climate system (see Section 5.7) has to be considered. Although the size of the tropical Pacific means that it exerts a greater influence on the global climate than these other variations, this does not mean they cannot be the dominant factor in fluctuations in their own backyard. What is more, it is possible that these other 'oscillations' can gang up to alter the otherwise apparently predictable consequences of an ENSO event. The recent work of William Lau and colleagues[29] in respect of predicting precipitation in the United States is a good example of this phenomenon (see Section 5.3). So, in considering the prediction of quasi-cyclic behaviour in the climate we cannot focus solely on one apparently dominant component of the interaction of the atmosphere and the oceans, but must take a global view.

5.7 Other interannual oscillations

The impact of the major El Niño events of 1982/83 and 1997/98 stimulated the meteorological community to look more closely at possible quasi-periodic atmosphere–ocean interactions around the world. Initally these focused on the NAO (see Section 3.7), but increasingly they have assumed global proportions. Indeed, there is now a concerted debate as to whether

[29] Lau, Kim & Shen (2002).

the NAO should be incorporated into a hemispheric *annular mode*, known as the Arctic Oscillation, that reflects the wider implications of circulation patterns around the the northern hemisphere.[30] Although this change has certain physical attractions, for historical reasons this book will stick with the NAO. This conservative approach also has the benefit of staying with a measure that is easily interpreted in terms of the nature of the circulation in the North Atlantic and also reflects the dominant part this aspect of the hemispheric circulation plays in the interannaul fluctuations of the climate at high latitudes of the northern hemisphere.

Although the NAO has attracted the most attention in recent years, perhaps the potentially most interesting of the new oscillations that have come to light in the last decade or so are in the Pacific Ocean, which covers nearly a third of the Earth's surface. This huge area, combined with its roughly symmetrical form (both latitudinal and longitudinal) makes it the most obvious region for longer-term coupled interdecadal variations of the atmosphere and ocean. While shorter-term fluctuations in the 2- to 8-year range are dominated by the ENSO variation, interdecadal changes have the largest amplitude in the North Pacific rather than in the eastern tropical Pacific. In addition, there is a coherent pattern of surface temperature variability in the southern hemisphere with cold and warm anomalies in the region of New Zealand alternating in opposition with warm and cold anomalies in the south-eastern tropical Pacific.

In the North Pacific these fluctuations, although being closely allied to the North Pacific Oscillation (NPO) identified by Sir Gilbert Walker in the 1930s (see Section 3.7), have become known as the Pacific Decadal Oscillation (PDO). A measure of this phenomena has been developed by Nate Mantua and colleagues at the University of Washington, Seattle, that is a measure of SST anomalies in the North Pacific: *warm events* are defined as when the temperature of the north-east Pacific and much of the tropical Pacific is above normal while much of the eastern Pacific north of 20° N is below normal, and atmospheric pressure is below normal over the Aleutian Islands.[31] During *cool events* the reverse conditions apply. Typical PDO events have persisted for 20 to 30 years. Their climatic impact is most evident in the North Pacific/North American sector, while secondary signatures exist in the tropics, in contrast to the dominantly tropical nature

---

[30] The case for analysing global mid- and high-latitude circulation in terms of annular modes is made in the review papers Thompson & Wallace (2000) and Thompson, Wallace & Hegerl (2000). The arguments for sticking to the NAO rather than moving to the Arctic Oscillation are set out in Ambaum, Hoskins & Stephenson (2001).

[31] Mantua *et al.* (1997).

of ENSO. During the twentieth century cool events prevailed from 1890 to 1924 and again from 1947 to 1976, while warm events dominated from 1925 to 1946, and from 1977 until around 1998 there appears to have been a move back towards cool PDO conditions.

The low-frequency variability over the North Pacific in wintertime is closely linked to the intensity of the semi-permanent low pressure system near the Aleutian Islands (the *Aleutian low*). During warm events the Aleutian low is deeper than normal and the westerly winds across the central North Pacific strengthen. This leads to above normal surface air temperatures over much of north-western North America and below normal precipitation, while over the south-eastern United States the reverse is the case. During cool events the opposite patterns occur more often over North America. This pattern also has hemispheric implications. When the warm PDO conditions are in phase with a positive NAO they reinforce the strength of the hemispheric circulation and hence to the global warming, as happened during the 1980s and 1990s.

In addition, there is a link with ENSO behaviour, as the cool phase of the PDO tends to coincide with periods when there are an above average number of La Niña events, whereas warm conditions coincide with more frequent El Niño events. So, the combination of PDO and ENSO signals can either reinforce or cancel out patterns over North America. For instance, the cool phase PDO and La Niña will act in concert, as will the warm phase PDO and El Niño. The reverse combination will tend to negate one another. So, the phase of the PDO appears to modulate the strength of ENSO events.

To the extent that there is confusion between the NPO and the PDO, it is compounded by the fact that interdecadal fluctuations in ENSO and the southern hemisphere are known as the Interdecadal Pacific Oscillation (IPO). It is to be hoped that the meteorological community will come to an early agreement on nomenclature, but, in the meantime, we will work with the terms used by individual research groups. When the PDO is in a warm phase the IPO is defined as being in a 'positive' phase, which coincides with the south-eastern tropical Pacific being warm.

In the tropical south-western Pacific there is a region of lower surface air pressure, where converging rising air produces cloud and rainfall, which appears also to be influenced by these changes. Known as the South Pacific Convergence Zone (SPCZ), it runs diagonally south-east from the Solomon Islands to Samoa and beyond. While it usually shifts little during the year, its position is linked to ENSO variations. During El Niño events it is displaced east, and during La Niña events west, of its mean position. On the longer term the zone has shown a striking eastward displacement since 1977, compared with the period 1948 to 1976. The two phases of the IPO also appear to modulate year-to-year ENSO precipitation variability over Australia. The positive phase enhances the prevailing west to south-west atmospheric circulation in the region, and the negative phase weakens this circulation. Farther east, annual temperature patterns over New Zealand show a significant 14- to 18-year periodicity that appears to be closely associated with southern hemisphere SST anomalies.[32] This periodicity does not, however, appear in the spectrum of tropical Pacific SSTs, where the greatest coherence with New Zealand temperature anomalies occurs in the 5- to 10-year periodicity range.

Work recently published by Yves Tourre and colleagues at the International Research Institute for Climate Prediction (IRI) of Columbia University, New York, together with Warren White of the Scripps Institution of Oceanography has helped clarify the overall position.[33] This analysis of sea level pressure and SST patterns of the Pacific Ocean between 1900 and 1991 has identified a significant distinction between decadal and interdecadal fluctuations. Centred on 16.7 years and 11.2 years, these two periodicities both show a broad swing back and forth between above average SSTs in mid latitudes and below normal values in the tropics, and the reverse conditions at the opposite end of the oscillation. These oscillations are spatially coherent, and are symmetric about the equator.

The case for a spatially coherent set of decadal and interdecadal interactions throughout the Pacific, and establishing a standardised nomenclature for the various oscillations, is reinforced by proxy data. Studies of a 271-year record of the variability of the Sr/Ca ratio in a coral from the island of Rarotonga, which is located at 21.5° S and 159.5° W in the region of the eastern SPCZ, underline these connections.[34] The record, dating back to 1726, shows a distinct pattern of decadal variability, with repeated decadal and interdecadal SST regime shifts greater than 0.75 °C. Comparison with the PDO indicates that several of the largest decadal-scale SST variations at Rarotonga are coherent with shifts in SSTs in the North Pacific. These observations may also be consistent with tree-ring studies carried out in Southern and Baja California, extending back to 1661 and showing clear evidence of a bidecadal oscillation that appears to be due to the PDO.[35] This all fits in with a hemispheric symmetry that suggests tropical forcing may be an important factor in at least some of the decadal variability observed in the Pacific Ocean.

Variability of the extratropical southern hemisphere circulation on timescales of a few months to a few decades is also linked to how tropical

[32] Folland & Salinger (1997).   [33] Tourre *et al.* (2001).
[34] Linsley, Wellington & Schrag (2000).   [35] Biondi, Gershunov & Cayan (2001).

convection is modulated by the intraseasonal oscillations in the tropics (see Section 5.8) and by ENSO. In mid latitudes these circulation changes affect the frequency of winter storms and rain to central Chile and east of the Andes. Farther south, SST anomalies tend to move eastward in the general flow of the Antarctic Circumpolar Current, suggesting a coupling in the ocean–atmosphere system of the region. This has become known as the Antarctic Circumpolar Wave.[36] Four huge pools of alternating above and below normal temperature water appear to circulate every eight years or so, and appear to be linked with changes in rainfall patterns over the southern continents. In addition, there are now sufficient data to establish the existence of an annular mode in the pressure patterns at high latitudes of the southern hemisphere. Known as the Antarctic Oscillation (AAO), it can be regarded as mirroring the Arctic Oscillation, and its better-known components, the NAO and PDO, in the northern hemisphere.[37] Indeed, as long ago as 1928, Sir Gilbert Walker[38] noted that 'Just as in the North Atlantic there is a pressure opposition between the Azores and Iceland, . . . there is an opposition between the high pressure belt across Chile and the Argentine on the one hand, and the low pressure area of the Weddell Sea and Bellingshausen Sea on the other'. It was not until the 1980s that sufficient data became available establish the reality of this oscillation.

In the southern oceans, the effects of ENSO extend to high latitudes.[39] In particular, this influence is seen in the distribution of sea ice coverage. Although the total extent has not changed significantly over the last 20 years (see Section 5.3), ENSO events appear to affect regional ice distributions. A study of satellite data from 1982 to 1999 identified changes in sea ice cover around Antarctica.[40] The strongest links were observed to be in the Amundsen, Bellingshausen and Weddell Seas. Within these sectors, higher sea level pressure, warmer air temperature and warmer sea surface temperature are generally associated with the El Niño phase. This reduces sea ice concentration in the Ross Sea, and shortening of the ice season in the eastern Ross, Amundsen and far western Weddell Seas. Four El Niño episodes over the 17-year period occurred at the same time as ice cover retreats in the Bellingshausen and Amundsen Seas, showing a close link between ENSO and this region of the Antarctic.

Long-term fluctuations in the behaviour of the Atlantic and Indian Oceans exert a major influence on weather patterns in Africa, India and

[36] White & Peterson (1996).

[37] Although the concept of the Antarctic Oscillation had been explored in earlier papers, its definition appears in a more accessible form in Gong & Wang (1999).

[38] Walker (1928).   [39] Peterson & White (1998).   [40] Kwok & Comiso (2002).

South America. In some instances these effects can cancel out the influence of ENSO events. In particular, a pattern of variations in the Indian Ocean seems to play an important part in the variety of responses to ENSO phases that occur in southern Africa and the monsoon in India. Unlike the Pacific Ocean, the surface wind flow over the tropical Indian Ocean does not have a prevailing easterly component. Strong upwelling off the Horn of Africa during the summer monsoon provides seasonal cooling of SSTs and reinforces the surface pressure patterns that drive the monsoon winds of India. Recent analysis of historical ocean and atmospheric data suggests that every few years or so there is an east to west oscillation of warm waters similar to El Niño and La Niña events of the Pacific.[41] In 1997, in association with El Niño, temperatures in the western Indian Ocean were well above average while in the east they fell more than 2 °C below normal. This triggered the wettest year on record in East Africa, while southern Africa had considerably more rainfall than was expected given the strength of El Niño event in the Pacific. At the same time, dryer conditions occurred over Indonesia and northern Australia. This behaviour may also help to explain why the strength of the monsoon over India responded sometimes unexpectedly to El Niño and La Niña events in the Pacific.

The fact that this pattern of SSTs in the Indian Ocean was, however, termed a 'dipole' has led to a lively debate as to whether using such terminology is leading to needless confusion.[42] The nub of the problem is whether SSTs in the Indian Ocean do indeed see-saw between the eastern and western extremities and, if so, whether they can be regarded as separate from ENSO activities. The case for the SST dipole does not appear particularly convincing,[43] and given the extent of the Southern Oscillation (see Fig. 3.7) it is probably not helpful to seek to decouple events in the two ocean basins. So, as with the plethora of 'oscillations' in the Pacific Ocean, seeking to establish a separate nomenclature does not shed additional light on an already murky area. This need not mean, however, that SST anomalies in the Indian Ocean do not have a major impact on rainfall patterns on the adjacent land masses.

The same issues have arisen in the case of the Atlantic Ocean. Because this ocean has the characteristics of two large basins linked at the equator, rather than of a single basin, it is likely to exhibit somewhat

[41] These patterns have been identified in both the tropical region of the Indian Ocean (see Saji *et al.* (1999), Webster *et al.* (1999)) and in the southern subtropical zone (Behera & Yamagata (2001).

[42] Hastenrath (2002).

[43] See for example Hastenrath & Heller (1977), Lamb (1978) and Hastenrath (2002).

different interannual behaviour to the Pacific and Indian Oceans. Studies in the 1970s established that rainfall in both the Sahel and north-east Brazil was more strongly correlated with the temperature contrast between the two hemispheres in the tropical Atlantic. The issue of a 'dipole' emerged in the 1990s, notably in a paper published in 1996 of studies of temperature anomalies since the 1960s, which presented evidence of SST patterns in the tropical Atlantic exhibiting a dipole pattern, the foci of which are centred on the tradewind zones at about 15° N and 15° S. Positive and negative SST anomalies appeared to oscillate in phase with a period ranging from 10 to 20 years.[44] Subsequent analyses of a wider range of meteorological data during the twentieth century have provided a less clear-cut picture, with the patterns of decadal temperature fluctuations on either side of the equator showing little evidence of a significant correlation.[45] The SST variability is strongly correlated with wind stress anomalies in the trade wind zones. The decadal variation of SSTs in the tropical Atlantic does, however, appear to be a longstanding feature of the region. Sediment cores from Cariaco Basin off the coast of Venezuela covering the last eight centuries show a highly significant periodicity of about 13 years.[46]

In the North Atlantic, interannual and decadal fluctuations are dominated by the NAO (see Section 3.7). In addition, there is evidence of a tripole pattern.[47] This pattern is most obvious during winter. When SSTs are below normal in the subpolar region, a warm anomaly occurs in the middle latitudes centered off Cape Hatteras, and a cold subtropical anomaly occurs between the equator and 30° N. Conversely, when subpolar regions are warmer than normal, the middle region has below average temperatures and the tropics are above normal. There is, however, considerable debate over the extent to which the NAO, the tripole pattern and the interhemispheric oscillation are part of a coherent response of the Atlantic Ocean as a whole.

Perhaps the most interesting feature of the extended web of interactions affecting the tropical Atlantic is a decadal 'preference' in SST variability and similar periodicities in both the Pacific and Indian Oceans (see Section 4.4). For instance, coral records for Palmyra Island in the central tropical Pacific show a 12- to 13-year periodicity that is highly coherent with long equatorial Atlantic and Indian Ocean climate records, implying a unified

---

[44] Nobre & Shukla (1996).

[45] See for example Rajagopalan, Kushnir & Torre (1998), Dommenget & Latif (1999) and Ruiz-Barradas, Carton & Nigam (2000).

[46] Black *et al.* (1999).   [47] Sutton & Allen (1997).

phenomenon.[48] The possibility of close links with the Pacific has been the subject of considerable work. The narrow strip of Central America means there is an 'atmospheric bridge' between the two ocean basins in tropical and subtropical regions. This link is reflected in both wind patterns and the flux of fresh water, in the form of rainfall. Modelling studies suggest that in terms of decadal fluctuations the North Atlantic can lag behind events in the North Pacific by 2 to 4 years.[49] In respect of the freshwater flux the connection is likely to be more complex.[50] Using the ECHAM4 global climate model, Mojib Latif, of the Max-Planck-Institut für Meteorologie, Hamburg, obtained an out-of-phase relationship between decadal fluctuations in the eastern tropical Pacific SST and the freshwater flux over the tropical Atlantic. This result is supported by a marked increase in sea surface salinity at Bermuda that parallels the model's prediction of changes in the difference between precipitation and evaporation in tropical Atlantic. The potential importance of this prediction is that the salinity of the tropical Atlantic will affect the strength of poleward oceanic circulation (*thermohaline circulation*, see Section 5.9). This might, in due course, affect the NAO, and statistical analysis of SSTs in the eastern tropical Pacific and the NAO shows a significant correlation between the two with a 30-year lag. So some of the impact of events in the tropical Pacific may take around 30 years to have a measureable impact on the North Atlantic.

These lengthy delays also link in with the other aspect of North Atlantic variability identified in Section 3.7 – the Atlantic Multidecadal Oscillation (AMO). This 65- to 80-year fluctuation has not only shown up in meteorological and proxy records, but has also been a feature of certain GCMs, notably the one developed at the Geophysical Fluid Dynamics Laboratory (GFDL), at Princeton.[51] Although this multidecadal oscillation appears to be driven by variations in the surface heat flux, and hence by atmospheric variability, it appears to be coherent with parallel fluctuations in the strength of the thermohaline circulation of the North Atlantic (see Section 5.9).

An overarching feature of this growing interest in oscillations around the world is the extent to which they interact with one another. Over and above the links already noted in respect of the links between the tropics and higher latitudes in both the Atlantic and Pacific Oceans, there is growing evidence of coherent connections between different ocean basins. This intricate set of connections further complicates matters, especially where

[48] Cobb, Charles & Hunter (2001).   [49] Gallego & Cessi (2001).   [50] Latif (2001).
[51] See for example Manabe & Stouffer (1988), Delworth (1996) and Delworth, Manabe & Stouffer (1997).

there may be an appreciable lag between events in one basin and those in another. More sophisticated models will play a part in unravelling the nature of these connections. Improved understanding will, however, depend in the first instance on better measurements of the properties of each ocean basin over time.

5.8 Intraseasonal oscillations

The 30- to 60-day periodicities in cloudiness and rainfall in the tropics, known as the Madden–Julian Oscillation (MJO) or intraseasonal oscillations (see Section 3.11), have become an important part of understanding both ENSO and other oscillatory behaviour in the climate.[52] They appear to involve a series of feedback processes in the tropical atmosphere that reflect some of the features of the processes at work in the tropical Pacific during El Niño. Indeed, these shorter-term fluctuations may play an important part in the strength of El Niño, as they appear to modulate the strength of the atmosphere–ocean interactions that drive these events. Depending on whether these waves of cloudiness reinforce or counteract the processes in the growth or decay of ENSO events, they play an important part in the overall scale of an event.

To understand how these two meteorological phenomena may be linked we must consider a few basic features of tropical meteorology. The first is the well-recognised circulation process that underlies tropical weather. Known as the 'Hadley cell', this involves moist air heated by the Sun rising close to the equator to produce a region of towering clouds. The moisture in the rising air condenses out, releasing latent heat. The resulting cold dry air at a height of 15 to 20 km spreads out polewards and then descends around 20° north and south. This dry air is the reason why the major deserts are located in these latitude bands. This air then returns towards the equator picking up moisture over the tropical oceans, before the process starts all over again. This surface flow defines the equatorial region of cloudiness: the ITCZ (see Section 3.11). This process continues throughout the year but the position of the ITCZ shifts to reflect in part the annual motion north and south of the overhead sun (see Section 5.4).

The other atmospheric phenomenon is 'Kelvin waves'. These are the atmospheric equivalent of the waves observed in the surface waters of the tropical Pacific (see Section 5.4). These waves are observed in the

---

[52] Madden & Julian (1971), (1972) and (1994).

stratosphere as an eastward-moving pressure field, where, as we will see in Section 7.3, they are part of the explanation of the QBO in equatorial stratospheric winds. But in the troposphere they had not been observed, as they were obscured by the natural variability of the weather until satellite observations unscrambled the picture. Moreover, according to simple theoretical physical analysis they should move much more rapidly, having a period of around 10 days. The explanation for the slower movement appears to be that the release of latent heat in the ITCZ has the effect of slowing the propagation of the Kelvin waves. In effect, the more rain that is formed in the Hadley cell the slower the waves move.

The importance of the release of latent heat in the rising air of the Hadley cell to control the movement of Kelvin waves may be the key to our understanding. But the situation is not simple; although it may well be that the observed periodicity is an essential feature of the Hadley cell. In effect, the pace of circulation may be governed by the rate at which the returning low-level air can pick up moisture. GCMs tend to predict that there is an optimum speed of circulation. The more moisture that is fed into the tropical 'boiler' the faster the cell will circulate. But above a certain speed the time of the return leg is cut too short to pick up enough moisture: a negative feedback that reduces the energy input to the cycle and slows it down. This means the Hadley cell has a characteristic circulation time of around 45 days. Moreover, the natural variability of the tropical weather is bound to disturb the cycle and so it will fluctuate appreciably around this natural period.

The essentially chaotic nature of weather events in the tropics makes the precise behaviour of the MJO difficult to predict: it can be strong in some years and almost absent in others. This is frustrating as it is at its strongest from September to May, and hence can play a major part in the timing of ENSO events. So, if one of these bouts of activity, with its strong convection and anomalous westerly winds, hits the western Pacific just as an El Niño event is ready to hatch, it can stimulate its rapid development. This is what appears to have happened in early 1997. But, what triggers a bout of activity is not fully understood. Furthermore, there is a 'chicken and egg' situation in that intraseasonal oscillations often exhibit a strong relationship to the phase of the ENSO cycle. Overall, MJO activity tends to be weak or absent during moderate or strong El Niño episodes. In contrast, it is often substantial during ENSO-neutral years and during weak La Niña episodes. This means that the MJO is inclined to exert a negative feedback on the switch between ENSO states.

Attempts to improve predictions of the impact of the MJO on ENSO may rest on overcoming two challenges. The first is to provide the correct mathematical treatment of tropical convective (rainfall) processes, which is currently missing in GCMs. The second challenge is to establish whether there is a physical link resulting from the fact that the periodicity of the MJO almost matches the time taken for oceanic Kelvin waves (see Section 5.5) to travel across the Pacific Basin. Analysis suggests that in certain circumstances there can be a near-resonant response between the lower frequency components of the MJO and variability of the depth of the thermocline that is driven by Kelvin waves.[53]

In the meantime, in spite of the roughly periodic nature of the MJO and the sense of the underlying orderly phenomenon, current forecasts can only be made on the basis that, if the last two or three cycles were, say, 45 days apart, it is reasonable to predict the next one will come along in about 45 days. This empirical approach has, however, not been particularly successful. Once an active area is identified, however, satellite imagery can be used to predict the short-term future movement of the active regions, but this is hardly an adequate input for seasonal forecasts of the type described in Section 5.6. In the shorter term satellite-derived data are used to detect regions where convective activity is abnormal. These departures from normal are a fundamental guide to monitoring and predicting the progress of the MJO as it propagates around the global tropics.

The challenge of the chaotic aspects of the MJO to seasonal forecasting extends to higher latitudes. Because the MJO affects the distribution and intensity if tropical rainfall, it influences the position and intensity of the subtropical high pressure regions and the mid-latitude jet streams. As a result, in winter it has an important impact on storminess and temperatures over North America. More generally, ENSO appears to have some influence on the amount of blocking from year to year. In the northern hemisphere over the eastern Atlantic, during the cooler seasons, blocking is more common when there are La Niña conditions in the tropical Pacific, a fact that may be linked to stronger MJO activity during these events.

During the summer, the MJO combines with the ENSO cycle to modulate hurricane activity in both the Pacific and the Atlantic basins. In addition, its phase affects tropical storm development over the tropical and subtropical North Pacific and North Atlantic. For example, MJO-related descending motion over the tropical North Atlantic suppresses tropical storm development, whereas MJO-related ascending motion is quite favourable for tropical storm development.

[53] Hendon, Liebmann & Glick (1998).

The way forward will lie in developing a better understanding of how the MJO fits in with longer-term interactions between the atmosphere and the oceans. Current GCMs are capable of representing some features of the MJO, but tend to underestimate its strength.[54] Furthermore, in one case it has been possible to show that a model forced by the observed SST patterns displays a decadal timescale variability of MJO activity; a development that raises interesting prospects of understanding the role of the MJO in quasi-cyclic interannual fluctuations of the weather.[55]

5.9 The Great Ocean Conveyor

So far the consideration of the nature of atmosphere–ocean interaction has concentrated on the mixed layer of the oceans (i.e. the waters above the thermocline). Beneath this active surface zone, which is stirred by the winds, heated by the Sun, and altered by the passage of warmer or colder air and by the advection of warmer or colder water or the upwelling of cold water, there is a set of much more gradual but equally important motions at work. The process driving the deep waters of the oceans is *thermohaline circulation*. This results from changes in seawater density arising from variations in temperature and salinity. Where the water becomes denser than the deeper layers it can sink to great depths. The temperature depends on where the surface waters come from and how much heat the oceans either pick up or release to the atmosphere in both sensible heat and evaporative loss. The salinity of a given body of water depends on the balance between losses through evaporation as opposed to gains from either rainfall, or freshwater run-off from rivers and melting of the ice sheets of Antarctica and Greenland plus the pack ice of the polar oceans. In practice, there are few regions where sinking water has a major impact. *Deep water,* which is defined as water that sinks to middle levels of the major oceans, is formed only around the northern fringes of the Atlantic Ocean. *Bottom water,* which constitutes a colder denser layer below the deep water, is formed only in limited regions near the coast of Antarctica in the Weddell and Ross Seas.

Thermohaline circulation drives a worldwide pattern known as the Great Ocean Conveyor (GOC) (Fig. 5.12). This model, which was developed by Wallace Broecker, at the Lamont Doherty Laboratory, building on an earlier proposal by Henry Stommel, has had a profound impact on recent

---

[54] Slingo *et al.* (1996).   [55] Slingo *et al.* (1999).

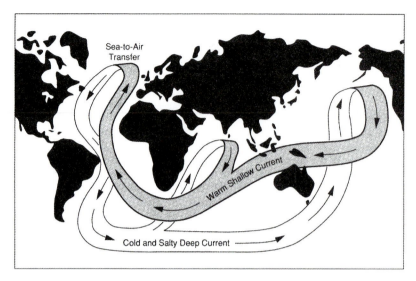

Fig. 5.12. The Great Ocean Conveyor – a schematic diagram depicting global thermohaline circulation. (After Trenberth, 1992, fig. 17.12.)

thinking about the possibility of rapid climate change.[56] Although it may turn out to be an oversimplified representation, it provides a useful way of considering the climatic implications of thermohaline circulation, and its implications for the climate. Because the GOC could exist in a number of very different states, it could be a fundamental factor in controlling climatic change. This is of particular relevance to the North Atlantic, where the current circulation carries the majority of heat to the Arctic. This warm water gives up its heat to the arctic air through evaporation. Then its low temperature and high salinity enables it to sink and form deep water that flows all the way to Antarctica. Here it is warmer and less dense than the locally frigid surface waters and so rises to become part of a strong vertical circulation process. Descending cold water from around Antarctica flows northwards into the Pacific and Indian Oceans, where there is no descending cold water. In the Atlantic this countercurrent is swallowed up by the much stronger southward flow.

In terms of possible involvement in cycles, the GOC could vary on a wide range of timescales, but it is in the millennial range that it may be of greatest interest. Broadly speaking it is estimated that the time taken for a packet of water to travel round the complete circuit is of the order of 1000 to 2000 years. So, it could be that significant changes in the climate

[56] See Stommel (1961) and Broecker (1994) and (1995).

are encapsulated in the circulation of the GOC and could show up on the millennial timescale. The fact that the 1450-year quasi-cycle (see Section 4.8) would fit in neatly with this fluctuation is an intriguing coincidence. Indeed, there is some evidence that there is a see-saw between changes in the Arctic and those in the Antarctic on these timescales.[57] In addition, the fact that this quasi-cycle has been linked to the occurrence of Dansgaard/Oeschger (DO) events in the North Atlantic touches on another important feature of the GOC. This is the possibility that its circulation could 'flip' from one state to another, bringing about a sudden change in regional climate.[58] So there is good reason for including current thinking about the dynamics of the global ocean circulation patterns when considering the origin of cycles on the centennial and millennial timescale.

5.10 Summary

This brief review of how some components of the global climate system may interact in a quasi-cyclic manner provides the starting point for trying to make sense of observed weather fluctuations. It indicates that in terms of climatic autovariance understanding the behaviour of the tropics is probably the starting point to explaining apparently regular variations in the weather, with particular emphasis on ENSO. The growing evidence that tropical oscillations appear in the Indian and Atlantic Oceans has, however, complicated matters. Furthermore, the fact that fluctuations at higher latitudes in the Pacific and the Atlantic represent a significant part of the global variance means that extratropical behaviour is an integral part of prolonged quasi-cyclic behaviour.

Progress in understanding the seething mass of oscillations will depend crucially on improved models of ocean–atmosphere interactions. This will involve not only better representations of the dynamics of separate ocean basins but also unravelling the subtle links that exist between separate basins. These may well go in and out of phase over time, leading to the amplitude of quasi-oscillations varying dramatically over time. In addition, in extending this insight to higher latitudes, agreement must be reached on the cause of the enigmatic and ubiquitous QBO, both in the stratosphere, and, more importantly, in the troposphere. Without this fundamental piece of the jigsaw it may be impossible to make sense of the whole range of quasi-cyclic fluctuations that so clearly demonstrate in the global climate

---

[57] Broecker (1998) and Blunier & Brook (2001).   [58] Broecker (1997).

that everything is connected to everything else. Furthermore, without this basic understanding it is difficult to consider the additional problem of external influences, whose periodicities may be close to those of the natural quasi-cyclic fluctuations of the global climate system. None the less, even though this basic insight is only starting to emerge from current research, we must now turn to the extraterrestrial factors that inevitably complicate the picture further.

Chapter 6

Extraterrestrial influences

Therefore the moon, the governess of floods,
Pale in her anger, washes the air,
That rheumatic diseases do abound:
And through this distemperance we see
The seasons alter: hoary-headed frosts
Fall in the fresh lap of the crimson rose.

Shakespeare (*A Midsummer Night's Dream*)

The Earth's weather cannot be considered in isolation. Thus far we have considered how the natural variability of the global climate could lead to quasi-cyclic changes over periods greater than a year. In so doing, it could be assumed that the Earth's tilt was constant and its orbit around the Sun was identical each year. Also it could be assumed that the energy output of the Sun is constant and that no other forces are present to perturb this orderly picture. None of this is correct. So we now have to consider how the natural autovariance of the climate may be affected by these external influences.

Three principal perturbations need to be addressed. First, there is the evidence that the Sun's output varies with time and that this is cyclic. Second, there are the tidal forces that act on the Earth due to the properties of the Moon's orbit and the more distant influence of the motion of the planets. Third, on the much longer timescale there are the periodic changes in the Earth's orbital parameters. The difference in timescale is important. Observations of periodicities in the Sun's behaviour only extend to a few hundred years, although they may well occur on longer timescales. Similarly, small variations in tidal forces that may be important in terms of this book are measured in tens and hundreds of years. Moreover, there are physical arguments to suggest that the observed fluctuations in the Sun's output may, in part, be linked to the forces exerted by the motion of the planets. So there is good reason to explore the evidence of these two extraterrestrial influences together. By way of contrast, variations in the Earth's orbital parameters are measured in tens and hundreds of thousands of years. So their effect on the climate can be considered separately and are of importance solely in discussing the origins of the ice ages.

The possibility of a link between solar variability and tidal forces needs to be borne in mind from the outset. Because the influence of the motions of the planets on both the Sun and the Earth's atmosphere will involve the same periodicities (though they will not be of the same physical magnitude), it will be difficult to unscramble the two effects when examining the meteorological records. None the less, it is easier to start by taking the solar and tidal variations in turn. Once we have identified the important features of the two sources of variability in the weather we can then look at the arguments for their being connected.

6.1   Solar variability: sunspots, faculae and coronal holes

When compared with the Earth's climate, the behaviour of the Sun would seem to be a relatively simple matter. As a massive ball of ionised gas, it is comparatively homogeneous. It is, however, subject to colossal gravitational, electrical and magnetic forces. So it is a seething mass of convective and circulatory systems, which are all capable of oscillatory motions that produce a variety of features capable of having climatic consequences here on Earth. This means that its behaviour is probably every bit as complex as the Earth's climate, and we can only view it from afar.

There are three regions of the Sun that are of particular relevance to weather cycles. The first is the visible surface (the *photosphere*), which has a radiative temperature of around 5700 K and is the source of the bulk of the energy reaching the Earth. The energy output of this thin shell, some 100 km thick (the Sun's radius is 700 000 km), is affected by two principal features: transient dark areas known as *sunspots*, and brighter regions known as *faculae*.

The second region is an irregular layer above the photosphere, known as the *chromosphere*. The importance of this layer, as far as the Earth's weather is concerned, is that observations of its behaviour can provide valuable insights into the Sun's magnetic field. In addition, changes in the chromosphere are a measure of fluctuations in the ultraviolet (UV) and shorter wavelengths, which have a disproportionate impact on the properties of the Earth's upper atmosphere. Chromospheric measurements include studying the bright boundaries surrounding sunspots (known as *plages*), which are associated with concentrations of magnetic fields in these areas and provide a means of linking sunspot activity with solar magnetic fields.

The third region is the outer atmosphere of the Sun: the *corona*. This region is in many ways a mystery. Although tenuous, it has an effective temperature of around 1 000 000 K. Its shape changes with the sunspot cycle, and in terms of impact on the Earth's weather, its most important features are *coronal holes*. Often found near the Sun's poles, these dark regions were first detected by X-ray equipment on Skylab in the early 1970s, and have been monitored since. They cover about 20% of the Sun's area when the solar activity cycle is at a minimum, and, as activity increases, they are replaced by smaller-scale open field regions scattered over the solar surface. The total area of coronal holes is closely associated with the high-speed solar wind this appears to be a good physical proxy for the global scale 22-year solar magnetic cycle and shorter-term aspects of the modulation of the flux of cosmic rays entering the Earth's atmosphere.

Any discussion of solar variability must start with sunspots, as these have played a central role in the search for weather cycles. Not only are they of historic interest, but also they now seem to be implicated in some of the most convincing examples of regular fluctuations in the weather. So it is important to start by reviewing what is known about these enigmatic blemishes on the face of the Sun and how they could affect the weather here on Earth.

Since the seventeenth century, observations of the Sun's photosphere have revealed sunspots at lower latitudes, between $30°$ N and $30°$ S (Fig. 6.1). Over time they vary in number, size and duration. There may be up to 20 or 30 spots at any one time, and a single spot may be from 10^3 to 2×10^5 km in diameter with a life cycle from hours to months. Each sunspot consists of two regions: a dark central *umbra* at a temperature of around 3700 K and a surrounding lighter *penumbra* at about 4700 K. The darkness of sunspots is purely a matter of contrast. They appear dark compared with the general brightness of the Sun's brilliant 5700 K photospheric temperature. If they could be seen on their own they would appear blinding white.

The average number of sunspots and their mean area vary over time in a more-or-less regular manner with a period of about 11.2 years (Fig. 6.2).[1] During this fluctuation the rate of increase in their number exceeds the rate of decrease, with the increasing phase taking roughly 4 years and the declining phase taking about 7 years; the period varies between 7.5 and 16 years, and the amplitude varies by about $\pm 50\%$. Each cycle begins when the

[1] Sunspot numbers can be obtained from the NOAA website: ftp://ftp.ngdc.noaa.gov/ STP/SOLAR_DATA/SUNSPOT_NUMBERS. More varied information about solar activity is available from a wide variety of sources, and a good place to start is the World Data Center for the Sunspot Index: http://sidc.oma.be/index.php3.

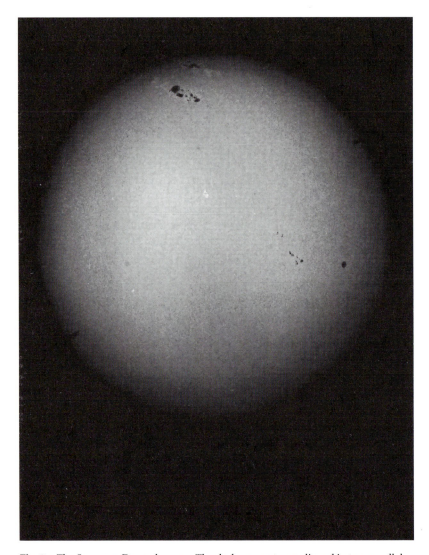

Fig. 6.1. The Sun on 21 December 1957. The dark sunspots are aligned in two parallel zones. (With permission of The Observatories of the Carnegie Institution of Washington.)

spots show up in both hemispheres some 35° away from the solar equator. As the cycle develops, the older spots fade away and new more numerous spots appear at lower latitudes (Fig. 6.3). Towards the end of each cycle, the number decreases and the spots are concentrated at latitudes some 5° from the equator. This cycle does not necessarily fall to nothing at the minima, because the new cycle will start at high latitudes before the old one has died away at low latitudes. The overlap can exceed two years.

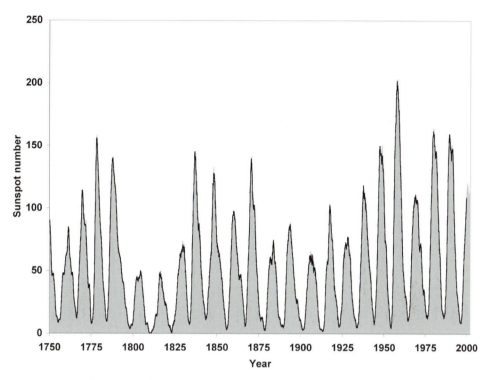

Fig. 6.2. The observed variations in the number of sunspots since 1750, showing the 13-month running mean of monthly numbers. (Data from the National Geophysical Data Center, NOAA: ftp://ftp.ngdc.noaa.gov/STP/SOLAR_DATA/ SUNSPOT_NUMBERS.)

The most frequently quoted measure of number of spots is the Wolf relative sunspot number, which is often referred to as the Zurich series on account of the original work being done by R. Wolf at the Zurich Observatory. This number is a measure of the number of groups of spots combined with the number of individual spots. It is normally calculated in terms of the monthly averages, and the standard curve of relative sunspot number consists of the 13-month running mean of the monthly figures. The variation in number during each 11-year period is more than two orders of magnitude greater than for any shorter period. It ranges from virtually no spots during the minimum of solar activity to just over 200 in the most active cycle that peaked in 1957. The instrumental record has now accumulated reliable data since around 1750 and is now in its twenty-third cycle of activity, which peaked at around 120 spots in 2000.

Faculae are closely associated with sunspots, and most easily observed near the solar limb (see Fig. 6.1). Their output is linked with the incidence of sunspots and it is now clear that they are a more important

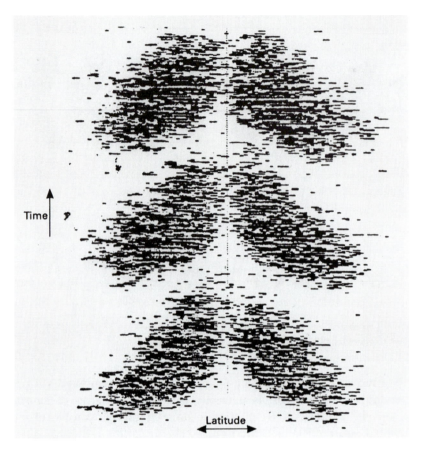

Fig. 6.3. The 'Maunder butterfly' showing the distribution of sunspots in heliographic latitude between 1874 and 1902, and the general movement in each hemisphere downward from immediately after the minimum in activity to immediately before the minimum. (From Mitton, 1977.)

factor in explaining how changes in solar activity could affect the weather, as their increased brightness is the dominant factor in changing solar output rather than sunspot darkening. In addition, the surface of the Sun is affected by a whole range of shorter-term disturbances, but in terms of their scale and duration they are of less consequence to weather variations from year to year. So for the purposes of this book we will concentrate on sunspots and their less-well-known cousins, faculae.

As described in Section 1.2 there have been many efforts to demonstrate that sunspots affect the amount of energy reaching the Earth from the Sun. Ground-based observations were, however, unable to provide convincing evidence as perturbations due to variations in atmospheric absorption swamped any small changes in the Sun's output. The advent of

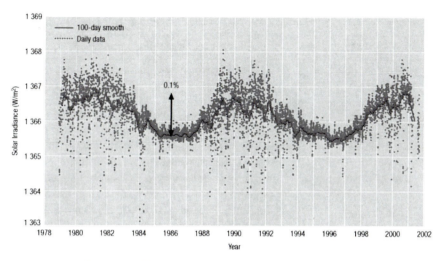

Fig. 6.4. Satellite measurements since 1978 confirm that the total output of solar energy has varied in line with the number of sunspots (with permission of WMO).

accurate satellite-borne instruments has, however, changed the situation radically. Starting with the launch of the satellite Solar Maximum Mission (SMM) in 1980, a series of satellites have made measurements that have produced unequivocal observations of how the Sun's output varies with the 11-year cycle in solar activity (Fig. 6.4).[2] What these results show is that the level of total solar irradiance (TSI) is greatest around the solar maxima and is essentially the same for the two observed minima: the hint of rise from the first to the second minimum rise in TSI has been reinforced by the latest calibration of the satellite data, which suggests a rise of 0.05% between successive sunspot minima. This controversial result supports the notion that the sunspot number might not be an accurate measure of solar energy output. The amplitude of the solar cycle variation is slightly less than 0.1%. The variations on shorter timescales (days to months) are larger, being as much as a few tenths of a per cent.

The fact that the TSI rises with sunspot number was, initially, a complication. Because sunspots are areas of low solar luminosity, it was presumed that the output would decline with sunspot number, as the effect would be to block some of the Sun's output and so reduce the overall energy flux. Indeed, the initial of the first year of the SMM record appeared to support this

[2] For original analysis see Willson & Hudson (1991); the first trend analysis of sunspot minima appeared in Willson (1997), and the latest more controversial results in Willson & Mordvinov (2003). Up-to-date measurements of solar irradiance can be found on the NOAA website: http://www.ngdc.noaa.gov/stp/SOLAR/IRRADIANCE/irrad.html.

hypothesis, as sunspot darkening dominated fluctuations on timescales of days to weeks.[3] This led to a model that produced estimates of the effective area of sunspots and predicted that the higher the sunspot number the lower the TSI. Clearly, this simple approach did not explain the longer-term observed changes in solar output. Moreover, this assumed relationship with sunspot number would not fit the longstanding hypothesis that the cold period known as the Little Ice Age, and more particularly the colder weather of the seventeenth century, were the result of an almost complete absence of sunspots during this period, known as the 'Maunder Minimum' after the astronomer who brought this absence to the attention of the astronomical community.[4]

To explain why the reverse is observed, it is now accepted that an associated increase in the brightness of the faculae outweighs the effect of sunspots. Measurements of the area of the Sun that faculae cover have been made since 1874, but, on their own, these were not sufficiently reliable to produce estimates of past TSI. A solution has, however, been found in combining recent satellite UV measurements and ground-based observations of the 10.7 cm microwave flux from the Sun. While the satellite observations only go back to 1980, the microwave observations have been made since 1947 and provide an adequate measure to calibrate the radiative temperature of faculae back to 1874 with reasonable confidence.

Judith Lean, now at the Naval Research Laboratory, Washington DC, and P. Foukal of Cambridge Research and Information, Massachusetts,[5] first proposed a model that combined the changes in sunspot number and faculae brightness based on microwave observations. This showed that it was possible to obtain a good fit with satellite observations of the TSI using the fact that the total surface area of the bright faculae dominates variation in the Sun's output. This increases and decreases in phase with the solar cycle. In addition, it became clear that the short-time-scale variations superimposed on the 11-year cyclic variation arise from the emergence and evolution of plages and also from the modulation of the number of sunspots by the 27-day rotation period of the Sun. These effects are largest during periods of high activity.

The relationship between sunspot darkening and faculae brightening is not simple, as periods of highest sunspot frequency are not necessarily the times when the Sun is radiating the most energy. For instance, the Foukal and Lean model estimated that the Sun's radiance was higher at the peak of solar activity in 1980 than it was during 1959 when the sunspot number was higher. Nevertheless, this model explained the observed changes in solar

---

[3] Maunder (1922) and Eddy (1976).

[4] Eddy, Gilliland & Hoyt (1982).   [5] Foukal & Lean (1990) plus subsequent papers.

output since 1980, and supported the hypothesis that the colder weather of the seventeenth century was the result of the Maunder Minimum. The small change in the TSI (less than 0.1%) during the last two solar cycles is, however, on its own, an order of magnitude too small to explain observed correlation between solar activity and global temperature trends since the late nineteenth century. This has led to a number of models being developed to predict changes in TSI since the seventeenth century. These have drawn on various proxies for solar activity. They include the Sun's magnetic flux at the Earth (the aa geomagnetic index),[6] emission of singly ionised calcium (the Ca II K line at 393.4 nm that is a measure of magnetic activity in the chromosphere),[7] solar cycle length, solar cycle decay rate, solar rotation rate and various empirical combinations of all of these. The models produce a variety of interesting results because, unlike sunspot numbers, which return to essentially zero at each solar minimum, the other indicators, notably the aa index, show 11-year cycles imposed on a longer-term modulation.[8] Another terrestrial indicator of solar activity is the level of cosmogenic isotopes in tree rings and ice cores, notably ^{14}C and ^{10}Be, which also shows longer-term modulation.

An alternative approach has been to estimate the TSI in terms of the variation in brightness of the 'quiet' Sun based on observations of the behaviour of other Sun-like stars using the Ca II K line. Making the assumption that during the Maunder Minimum the Sun was in a quiescent 'non-cycling' state, various reconstructions have deduced widely varying values for the TSI during the Maunder Minimum relative to the present. Uncertainties in the assumptions made about the state of the Sun during that quiescent period produce a range of between 1 and 15 Wm^{-2} reduction in TSI compared with present mean values, although most estimates lie in the 2.5 to 5 Wm^{-2} range. Even so, this may be sufficient to explain much of the observed correlation between global climate change and solar variability during the last three to four centuries.

Judith Lean's latest analysis of the TSI (Fig. 6.5) estimates a 0.20% increase (2.8 W m^{-2}) between the Maunder Minimum and the mean level in the late twentieth century.[9] In addition, this work has calculated the variation of output over different spectral bands. Faculae dominate the solar irradiance variations at UV wavelengths: at 200 nanometres (nm) sunspots

---

[6] Monthly values of the aa index can be found on the NOAA website: ftp://ftp.ngdc.noaa.gov/STP/SOLAR_DATA/RELATED_INDICES/AA_INDEX/ Aa_month.

[7] Lean *et al.* (2001).

[8] Lockwood & Stamper (1999) and Lockwood, Stamper & Wild (1999).

[9] Lean (2000).

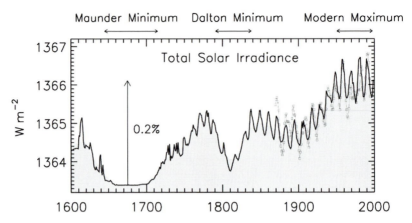

Fig. 6.5. An estimate of the annual total irradiance from the Sun since 1600 (from Lean, 2000). The shading identifies the 11-year running mean and the arrow shows the percentage increase to the mean of cycle 22 (1986 to 1996). The light grey line plus square symbols is the total irradiance (scaled by 0.999) determined independently by Lockwood & Stamper (1999). (With kind permission of J. Lean, Naval Research Laboratory, Washington DC.)

are a factor of two darker than they are at 500 nm, while faculae are two orders of magnitude brighter. Using satellite measurements and modelling work, Lean estimates that over the 11-year cycle solar radiance varies by 14% at around 150 nm and 8.3% at around 200 nm, but by less than 1% at wavelengths longer than 300 nm. What this means is that since the Maunder Minimum, solar irradiance is estimated to have increased by 0.7% in the wavelength range 120 to 400 nm, by 0.2% in the range 400 to 1000 nm, but by only 0.07% at longer wavelengths.

The fact that the TSI varies so much with wavelength is probably the key to the close statistical link between solar activity and observed changes in climate since the seventeenth century. So, we need to consider how this feature of solar variability could exert an undue influence on the weather by being amplified in the Earth's atmosphere. The significance of changes in the UV region is that these wavelengths are largely absorbed in the atmosphere and this could enhance the impact on the weather. In particular, wavelengths shorter than about 300 nm are strongly absorbed by oxygen and ozone in the stratosphere. This suggests a possible mechanism for solar variability being amplified in the stratosphere and hence influencing the weather at lower levels (see Section 6.3).

The nature of the Sun's magnetic field is also an essential part of understanding possible links between solar variability and the Earth's weather. In general terms, at the beginning of a sunspot cycle the solar magnetic field resembles a dipole that is aligned with the Sun's rotation axis. This means that at low latitudes the field lines are closed, whereas at higher latitudes

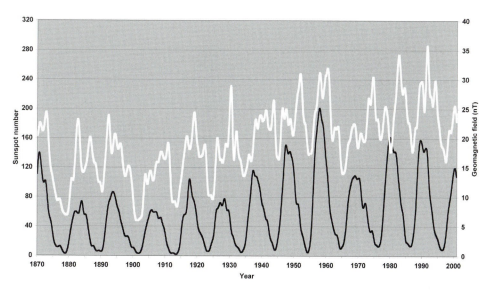

Fig. 6.6. Monthly sunspot numbers (black line) and the Earth's geomagnetic field (white line) (both smoothed with a 25-term weighted running mean). (Data from National Geophysical Data Center, NOAA.)

they are open. As the cycle builds up to a maximum this simple pattern breaks down into a disordered state. During the latter part of the cycle the dipole field is re-established.

At the detailed level, the magnetic polarity of sunspots, which has been observed since 1908, is observed to alternate between positive and negative in successive cycles. Sunspots tend to travel in pairs or groups of opposite polarity as if they are the ends of a horseshoe magnet poking through the surface of the Sun. During one 11-year cycle, as the spots traverse the face of the Sun in an east–west direction, the leading spots in each group in the northern hemisphere will generally have positive polarity while the trailing spots will be negative. In the southern hemisphere the reverse situation occurs with the leading spots being negative. It is this pattern that reverses between successive cycles. This 22-year magnetic cycle (the Hale cycle) could be the key to the amplification process. As noted in Chapter 3, the 20- to 22-year cycle has been more prevalent in climatic data than the more obvious 11-year sunspot cycle.

Equally interesting is that the Sun's magnetic field at the Earth (the aa geomagnetic index) rose by 130% between 1901 and around 1990.[10] While it exhibits periodic fluctuations apparently linked to the 11-year sunspot cycle, these are not precisely in phase with it, are less regular, and have proportionately less amplitude on the decadal timescale (Fig. 6.6). So, it is

[10] Lockwood & Stamper (1999) and Lockwood, Stamper & Wild (1999).

possible that a magnetic process could be identified which either plays a part in the observed rise in TSI between successive sunsport minima[11] or amplifies the impact of solar variability on the weather. If this is the case, then it could be that changes in the Sun over the last century have played a more important part in the observed global warming than in normally accepted.

The magnetic behaviour of the Sun has been explained in terms of an empirical dynamo model.[12] This model relies on the fact that because the Sun is gaseous, it does not rotate uniformly: bands of gases around the equator circle the solar axis once every 24 days or so, compared with over 30 days near the poles. At the same time, convection of hot gases from deep within the Sun's interior is balanced by the descent of cooler denser gases. Because of the high temperatures, the constituent atoms and molecules in the Sun's atmosphere are ionised and the motion of the charged gases generates magnetic fields.

The dynamo model involves the interaction of toroidal and poloidal magnetic fields generated by the Sun's surface differential rotation. It is proposed that because the Sun's surface rotates faster at the equator than at its poles and faster inside than at the surface, twisting motions affect the Sun's magnetic field in such a way that stresses gradually build up that are eventually released in a rash of sunspots and solar flares. In effect, those regions, which experience the strongest magnetic fields, build up a repulsion between the charged gases that make them lighter than in surrounding areas. This expansion cools the gas and so the regions of greatest magnetic fields are seen as rising cooler gases on the Sun's surface as the magnetic energy is dissipated. At this point, the field dies away and reverses polarity and builds up again to another peak, finally returning to its original configuration after 22 years.

The simple dynamo model does not, however, provide an explanation for the more complex features of the sunspot cycle and there is some doubt about whether it can be refined to do so. Furthermore, there are considerable uncertainties about the depth to which the convective processes extend and how these combine with the radiative processes at the surface. But for current purposes, we can assume that the differential rotation of the Sun is the key to the properties of both the sunspot cycle and the magnetic field properties, while recognising that future work on solar models may produce a rather different detailed picture of the processes involved.

[11] Willson (1997) and Willson & Mordvinov (2003).
[12] See, for instance the NASA, Marshall Solar Physics web page: http://science.msfc.nasa.gov/ssl/pad/solar/default.htm.

There is increasing evidence that the behaviour of the solar magnetic field has a direct and indirect effect on the climate. The direct effect is linked to how the fluctuations of the Sun's magnetic field affect the TSI and, in particular, variations in UV and shorter wavelengths. The indirect consequences of the solar magnetic field may relate to its influence on more energetic particles. First, the magnetic fields associated with sunspots and faculae affect the quantity of energetic particles emitted from the Sun. Second, the overall strength of the Sun's magnetic field alters the Earth's magnetic field and with it the amount of cosmic rays (energetic particles from both the Sun and from elsewhere in the universe) that are funnelled down into the atmosphere. How these magnetic effects may combine to alter the properties of the upper atmosphere and so influence the weather in a variety of ways will be examined in Section 6.3.

Thus far we have only considered the basic 11-year sunspot cycle and the 22-year Hale cycle. As is obvious from Fig. 6.2, the variation of the peaks in the successive cycles also shows evidence of a periodicity of around 90 years – often termed the Gleissberg cycle.[13] This cycle is also associated with the period between successive peaks in solar activity, which lengthens as the peak levels decline. Harmonic analysis of sunspot numbers over the period 1700 to 1986 shows that 90% of the variance is explained by 27 harmonics (Fig. 6.7a).[14] The most important peaks appear at 9.9, 11, 57 and 96 years. A comparable maximum entropy spectral analysis (MESA) study of sunspot data for the period 1700 to 1986 (Fig. 6.7b) produces a slightly different result, with the main peaks at 10.0, 11.1, 52.3 and 100 years. But when the analysis is split to consider periods before and after 1800, the position changes. It shows a splitting of the 11-year cycle before 1800 with a dominant 55-year cycle. After 1800 there is a single 11-year cycle and a 100-year peak. These periodicities are sensitive to the time interval over which the analysis is carried out. This implies that Fourier techniques are not useful for predicting the level of solar activity.

This conclusion is supported by an earlier MESA study of sunspot numbers.[15] When published in 1974, it confidently predicted that the twenty-first and twenty-second sunspot cycles would be of a much lower level of activity with 'twelve month running mean sunspot numbers greater than 100 not being observed again until approximately 2015'. In fact, both the subsequent cycles had maxima in excess of 150 and even the quieter twenty-third cycle peaked at 120 in 2000. This shows the pitfalls of attaching too much significance to the seemingly high-quality results of MESA. So, as with the weather, past sunspot cycles are not a reliable guide to future behaviour, in spite of the continued appearance of the 11-year cycle.

[13] Gleissberg (1958).   [14] Berger (1990).   [15] Cohen & Lintz (1974).

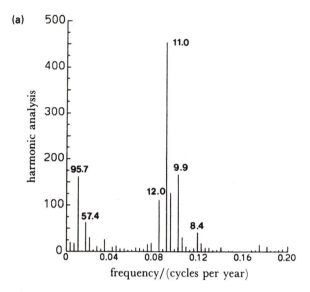

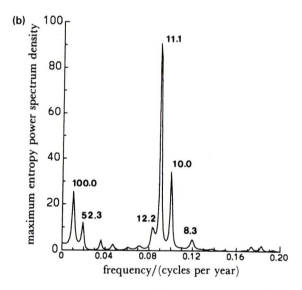

Fig. 6.7. The power spectrum of the sunspot time series from 1700 to 1986 calculated by (a) harmonic analysis and (b) MESA methods; (c) the 22-year magnetic cycle. (From Berger, Melice & van der Mersch 1990 with permission of the Royal Society.)

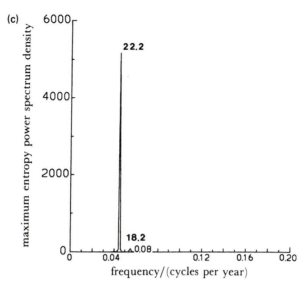

Fig. 6.7. (*cont.*)

A number of other interesting features emerge from the spectral analysis of sunspot numbers. First, the peaks at 11.1 and 10 years suggest the possible existence of a beat of around 200 years, which has been cited in support for the planetary theory of sunspots (Section 6.2). Second, the 90- to 100-year cycle is important, as this periodicity is frequently cited in Chapters 3 and 4. Furthermore, there is a marked parallelism between changes in mean temperature of the northern hemisphere and both sunspot numbers and the length of the sunspot cycle (Fig. 6.8). In particular, the close correlation between the length of the solar cycle and global temperature trends[16] sparked off an intense discussion that has continued in recent years. Moreover, if the sunspot cycle since 1700 is assumed to exhibit the same reversal of magnetic polarity that has been observed since 1908, the spectral analysis of the 22-year periodicity produces an exceedingly sharp peak (Fig. 6.7c).[17] This feature contains 62% of the total variance in the series and means that the magnetic cycle may be the most stable feature of solar behaviour.

In addition, two other interesting features emerge from spectral analysis of sunspot numbers. First, there are no significant periodicities between two and three years. So, there is no reason to believe that the QBO is of solar origin. Second, analysis of the record shows it does not behave as a Markov process (Appendix A.9), so the noise spectrum is 'white' (see Section 2.7).

---

[16] Friis-Christensen & Lassen (1991).   [17] Berger (1990).

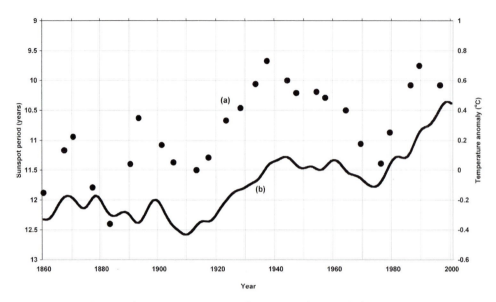

Fig. 6.8. The apparent association between (a) the period of the 11-year sunspot cycle between successive maxima and minima (three-period binomial running mean) and (b) the deviation of the average surface temperature of the northern hemisphere from the mean for 1951 to 1980 (21-year binomial running mean).

Here again, the inference is that the propensity of weather records to exhibit 'red' noise is not related to solar behaviour.

Returning to the longer periodicities, we have to use different data to find out more. For while there are many human observations, notably in China, of the incidence of sunspots prior to 1700, which can be used to infer that there is a 200-year cycle, these are so fragmentary that it is hard to reach reliable conclusions about longer periodicities. The best source of data about long-term solar variability prior to the eighteenth century is the level of cosmogenic isotopes, notably ^{14}C and ^{10}Be, in tree rings and ice cores. Variations in the concentrations of these isotopes are a consequence of changes in the Earth's magnetic field due to solar activity modulating the flux of cosmic rays entering the Earth's atmosphere, which controls the production of ^{14}C and ^{10}Be in the atmosphere. The usefulness of the radioactive isotope ^{14}C is that it has a relatively short half-life of 5730 years. Formed by the nuclear reaction between cosmic rays and nitrogen in the atmosphere, it is then taken up in carbon dioxide and is then absorbed by living plants in the process of photosynthesis. So tree rings, in addition to containing climatic data, can also provide information about the Sun's past behaviour. When solar activity is high, the magnetic shield is strong and the amount of ^{14}C formed, and hence incorporated in tree rings, is relatively low. Conversely, at times of low solar activity, the shield is weak, and more ^{14}C is formed. By measuring the amount of ^{14}C in tree-ring series, and comparing it with what would be present if it had been produced at a constant rate, it is possible to build up a measure of past solar activity.

There have been many studies of the spectra of ^{14}C fluctuations. Here we focus on the work of Minze Stuiver and Thomas Braziunas at the University of Washington, who analysed bristlecone tree-ring records of ^{14}C content going back 9600 years using MESA.[18] This work considered blocks of data averaged over 20 years to iron out shorter-term fluctuations. This means that it was able to look at periodicities longer than 40 years. It also removed the longer-term trend, which could not be satisfactorily explained, and so did not look at periods longer than 1000 years. It found impressive evidence of a fundamental period of 420 years, together with what looked like harmonics at 218, 143, 85 and 67 years (the fourth harmonic expected at 105 years was not present). The general conclusion is that these results provide clear evidence of a fundamental oscillatory mode of about 2.4×10^{-3} cpa (a 420-year period) together with several harmonics.

When the data were split into shorter chunks, the spectral properties alter to some extent. Notably, the last quarter of the record, running from 1570 BC to AD 1830, has a pronounced 127-year peak that is absent in other sections. This is interesting as this periodicity is close to the dominant feature in the lengthy rainfall records for China that cover most of this period (see Section 3.4). More generally, these results lend support for the periodicities of 114 and 208 years observed in the bristlecone pine tree-ring series (see Section 4.1). Furthermore, the variations over the last 1000 years show major lulls in solar activity around AD 1280, AD 1480 and AD 1680 (the Maunder Minimum).[19] These periods of low solar activity appear to coincide with periods of cooler climate in the northern hemisphere (Fig. 4.12).

International collaboration continues to refine the data on ^{14}C variations in tree rings and their associated timescale. The latest agreed calibration curve was published in 1998 (INTCAL98)[20] draws on a wide range of proxy data, including tree rings, coral and varves, to produce this new timescale. The important cycles in the fluctuations of ^{14}C in tree rings that can be attributed to solar activity are 206, 148, 126, 89, 26 and 10.4 years. While these figures are slightly different to those for periodicities identified earlier, they tell the same broad story of centennial solar variations, notably around 90 years and 200 years.

[18] Stuiver & Braziunas (1989).   [19] Stuiver & Quay (1980).   [20] Stuiver *et al.* (1998).

Different information can be obtained from the radioactive species ^{10}Be. This has a lifetime of 1.5 million years, which is useful for longer-term paleoclimatic studies. It is formed in relatively high quantities from collisions between cosmic rays and oxygen and nitrogen nuclei. Moreover, because it is not incorporated in an atmospheric gas, it is quickly precipitated out in rain or snow. Once in the soil or ice sheets it remains relatively immobile and provides an independent record of past solar activity. Analysis of ^{10}Be in Antarctic and Greenland ice cores provides confirmation of the principal periodicities in solar activity at around 90 and 200 years.[21]

In all the discussion of solar variability so far, we have concentrated on periodicities from two years upwards. There are, however, two more short-term periodicities to consider. The first is the most obvious, namely the rotation of the Sun. This could influence the weather on account of the variation of the radiance when there are large numbers of sunspots moving across the face of the Sun. At the same time, features in the structure of the Sun's magnetic field will sweep across the Earth and could affect the weather. The Sun's magnetic field normally has four sectors that alternately point predominantly towards and away from the Sun. Thin sheaths, across which the field direction reverses, known as *sector boundaries*, separate the sectors. As the Sun rotates, these sector boundaries sweep across the Earth at regular intervals.

Work over many years by the first Director of the National Center for Atmospheric Research, the late Walter O. Roberts, while at the High Altitude Observatory, Boulder, Colorado, explored the connection between this aspect of solar behaviour and the climate. Statistical analysis of the intensity of low-pressure systems in the northern hemisphere during winter showed a distinct pattern associated with the passage of sector boundaries.[22] A measure of this behaviour – the vorticity index – shows a pronounced minimum one day after sector boundary passage. In contrast, during other seasons there is no such link with sector boundaries. Although the significance of this observation has been the subject of considerable debate, it is important for two reasons. First, it shows that the Sun's magnetic field can have a direct, rapid and measurable influence on the lower atmosphere. Second, it suggests that the capacity of the Sun to affect the Earth's weather may vary with the seasons.

Although there is little evidence of cycles of around a month, there are two additional reasons for bearing in mind this feature of the Sun's possible influence on the weather. First, there is some evidence that changes in certain weather patterns can be linked to changes in the Sun's magnetic

---

[21] See for example Oeschger & Beer (1990) and Raisbeck *et al.* (1990).
[22] See for example Roberts & Olsen (1973), and Wilcox *et al.* (1979).

field, whatever the timescale. Second, for the reasons listed in Chapter 2, a periodicity of around 27 days could easily be misinterpreted when studying statistics. Not only is this period close to a calendar month, which is used to record so many weather data, but it is also close to the lunar month. While solar physicists avoid the statistical pitfall of aliasing by averaging solar parameters over this period, in other scientific areas there is a danger that weak effects due to the Sun's rotation may be attributed to other causes when sifting through the data.

The other interesting periodicity is a 155-day variation in the occurrence of high-energy solar flares. This feature has been analysed between 1874 and 1993 using the wavelet technique.[23] This periodicity is only evident around the maximum of solar activity in cycles 16 to 21. The periodicity started growing at cycle 16, peaked at cycle 19 and decreased in subsequent cycles, completely disappearing after cycle 21. While there is no accepted explanation for this behaviour, the fact that such a well-developed feature could appear over so many cycles provides an interesting additional insight into the quasi-cyclic properties of the Sun, and their potential to affect the Earth's atmosphere.

6.2 Tidal forces

The effects of the gravitational forces acting on the Earth as it orbits the Sun are complicated. The most obvious are the tides resulting from the combined pull of the Moon and the Sun. These tidal forces not only affect the movement of the oceans but also have a physical impact on the atmospheric motions. In addition, the forces exerted on the Earth by the changing position of the other planets will play a similar but much smaller role. These tidal forces also exert a stress on the Earth's crust, which could influence the release of tectonic energy, notably in the form of volcanic activity, and so conceivably modify the climate. Then there is the possibility of the same tidal forces due to planetary motions affecting the Sun's circulation and with it solar activity. Finally, there are the orbital effects of these motions. Because these can cause the Earth to speed up or slow down and also lead to small movements of the Sun about the centre of mass of the solar system, there is potential for small periodic influences on the Earth's climate.

Clearly, all these tidal effects are interlinked. But, as a first step, we need to consider each potential influence separately before trying to make observations as to what their combined effects may be. So, the obvious place

[23] Oliver, Ballester & Baudin (1998).

to start is with the semi-diurnal tides in the atmosphere and the oceans. These are caused by the gravitational attraction of both the Sun and the Moon. On the nearside of the Earth the atmosphere and the oceans are attracted towards both bodies, as is the Earth itself, which pulls it away from its fluid envelope on the far side. Because of the Earth's rotation and the Moon's orbital motion, any particular place on the Earth's surface experiences two complete cycles of high and low tidal stress every 25 hours. On average, the Sun's pull is roughly half that of the Moon. This means that there is a threefold variation between when the Sun and the Moon are on the same side of the Earth and pulling together, and when they are on opposite sides and so their effects partially cancel out.

The combined effects of the Sun and the Moon are complicated by the shape and period of both the Earth's orbit around the Sun and the Moon's orbit around the Earth.[24] The Earth's orbit is an ellipse with the Sun at one focus and has a period of 365.25 days (the plane of this orbit is known as the ecliptic). The Moon's orbit is also an ellipse with the Earth at one focus, and has a period of 27.533 days, but because of the Earth's motion around the Sun, the time between the conjunctions when the Moon, Earth and Sun are approximately in line, marked by the full Moon, is 29.531 days. These two periods are known respectively as the lunar 'synodic' and 'anomalistic' months. At the same time, because the Moon's orbit is at an angle of 5° 9' to the Earth's equatorial plane, the moves above and below this plane each month. It crosses the equator, when its declination to the ecliptic is zero, at a slightly different period of 13.661 days, and so the cycle length between equator crossings in the same direction is 27.332 days and is known as the 'tidal' month.

If the Moon completed an exact number of orbits of the Earth in the time it takes the Earth to complete a circuit of the Sun, the pattern of tidal forces would be relatively simple and reproducible. But this is not the case. The nearest it comes to this is that 13 tidal months amount to 355 days, which is sometimes referred to as the 'tidal year'. So although the Earth and the Moon return to approximately the same position after about a year, it takes much longer for more precise repetitions of alignment to occur.

This brings into play the relative motion of the Earth's perihelion – the point on its orbit when it is closest to the Sun – and the Moon's perigee – the point on its orbit when it is closest to the Earth. While the changing distance between these three bodies will exert a continual influence on the tidal forces, the relative positions of the perihelion and perigee play

[24] An accessible description of these astronomical facts is to be found in Lamb (1972), Chapter 6.

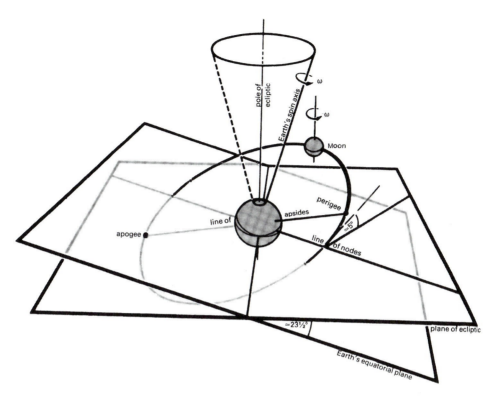

Fig. 6.9. Orbital geometry of the Earth–Moon system. The Earth's average orbital motion around the Sun defines the ecliptic and the Earth's spin axis rotates about the pole of the ecliptic once every 26 000 years because of precession. The Earth's equinox, the intersection of the equator and ecliptic, moves along the ecliptic at the same rate. The lunar orbit intercepts the ecliptic along the line of nodes that moves around the ecliptic because of the solar attraction. For the same reason the line of apsides precesses in the orbital plane. The Moon's spin axis remains normal to the ecliptic. (After Smith, 1982.)

an important part in the longer-term periodicities that may influence the weather.

There are two periods that matter most. The first is the 8.85-year period in the advance of the longitude of the Moon's perigee, which determines the alignment of the perigee with the Earth's perihelion (Fig. 6.9). The second is the 18.61-year period in the regression of the longitude of the node – the line joining the points where the Moon's orbit crosses the ecliptic. This period defines the exactness of the alignment of the Moon's perigee and the Earth's perihelion. There are two principal reasons to focus on these periodicities. First, they have values that, on the basis of the evidence presented in Chapters 3 and 4, seem to be linked to periodicities in weather statistics. Second, these motions appear to produce physical effects that could conceivably explain the observed fluctuations in the weather. But this selective approach must not ignore the various other periodicities that occur in these orbital motions, which could be significant, although there is less convincing evidence of this being the case.

The 18.61-year cycle is the most widely studied aspect of tidal stress. This is because the regression of the node defines how the angle of the Moon's orbit to the Earth's equatorial plane combines with or partially cancels out the tilt of the Earth's axis. This has the effect of altering the variation of the maximum declination of the Moon from the ecliptic. Because of the tilt of the Earth's axis, the equatorial plane is at an angle of 23° 27′ from the ecliptic. This combines with the angle of the Moon's orbit to the equatorial plane so that the maximum declination to the ecliptic is 28° 40′ N and S. But this extreme value occurs only every 18.6 years; at the opposite extreme, halfway between when the maximum declination is the difference between the tilt of the Earth's axis and the angle of the Moon's orbit from the equatorial plane, the value ranges only between 18° 21′ N and S. The significance of this variation is that when the declination is greatest, the tidal forces at high latitudes are greatest. Recent peaks in these forces have occurred in 1913, 1931, 1950, 1969 and mid-1988.

The importance of the 8.85-year cycle in the alignment of the Moon's perigee and the Earth's perihelion is not its direct impact on tidal forces, but how it combines with the 18.61-year cycle. Calculations of the tidal stress at high latitudes since AD 1100 show that there is a tidal resonance at about 179.3 years.[25] The significance of this period is that it is yet another candidate for the general group of periodicities that have been identified at around 180 to 200 years. Its physical significance will be considered on Section 6.3.

As noted earlier, there are a number of other periodic coincidences associated with the relative motions of the Earth, the Moon and the Sun. These include 13.6, 27.2 and 44 years. The first is the completion within $1^{1}/_{2}$ days of a whole number of 29.53-day synodic months (167), 27.55-anomalistic months (179) and 27 half years of 182.62 days in which the Earth moves exactly to the opposite of its orbit. The period of 27.2 years brings the corresponding near-repetition of the phase of the lunar month with the Earth in its original position in its orbit. The 44-year period coincides with the near coincidence between the completion of a whole number (45) of tidal years, consisting of 13 tidal months, and 44 calendar years. The importance of these periods is that they should lead to reproducible tidal forces around their completion. The fact that these periodicities are not prominent in the meteorological

---

[25] Roosen *et al.* (1976).

series reviewed in this book suggests that these particular tidal forces are not exerting a significant influence on the weather.

Looking beyond the tidal effects of the Sun, Moon and Earth, there is the influence of the other planets in the solar system. Of these, the motions of the giant planets Jupiter, Saturn, Uranus and Neptune are potentially the most interesting. These have orbital periods of 11.86, 29.5, 84 and 165 years respectively. So, singly or in combination they influence the tidal forces on the Earth. This possibility was the subject of considerable debate in the early 1970s, as it was suggested that these planetary motions could influence the release of tectonic energy.[26] In practice, because of its mass (318 times that of the Earth) and its relative proximity to the Earth, Jupiter is by far and away the most important factor in these planetary forces. Moreover, the fact that its period is close to the 11-year sunspot cycle means that it could either be a confusing factor in identifying the causes of periodicities in the weather or directly linked to the sunspot cycle itself.

In respect of the tidal influences of the planets on the Earth's atmosphere and oceans, detailed calculations show that the scale of these perturbations is small compared with the variations of the tidal forces due to the motions of the Earth and Moon around the Sun. So unless there is a good physical reason for the much smaller gravitational influences of the planets to have a proportionately greater impact, they are unlikely to have a significant effect on the weather. One such possibility is that although the giant planets' influence on the Earth's atmosphere and oceans may be trivial, they produce potentially important perturbations on the Earth's orbit. Work published Ren Zhenqiu of the Beijing Academy of Meteorological Science and Li Zhisen of the Beijing Astronomical Observatory examined the effect of the Earth being on one side of the Sun and all the other planets grouped in a tight arc on the other side, and whether this effect could be identified in terms of reproducible effects in Chinese climate records.[27] Known as a synod, this particular alignment of the planets occurs every 179 years or so, although every five or six cycles the interral can be as short as 140 years. The fundamental rhythm of these synods is defined by the approximate alignment of Jupiter, Saturn, Uranus and Neptune. The movements of Mercury, Venus, Earth and Mars define the time of year when the synod occurs.

The Chinese study examined those occasions when the remaining planets were grouped in an arc subtending less than 90° at the Earth. After 1600 BC in China, when the synod occurred in the summer half of the year, the subsequent few decades tended to feature warm summers. Conversely,

---

[26] Roosen *et al.* (1976).   [27] Gribbin (1982).

after winter synods there were more frequent cold winters. Moreover, where the grouping was narrower the effects tended to be more pronounced. The physical explanation of these observations appears to be the way the planetary configuration causes the Earth to speed up or slow down in its orbit. While the period of the orbit (365.24 days) remains unaltered, when the Earth is travelling towards the grouping of the giant planets it speeds up and when it is travelling away from them it slows down. This means it spends less time in the half of the orbit when it is closest to the conjunction of the planets and more time on the far side of the Sun. So if the synod occurs in the summer half of the year this period will be lengthened slightly and vice versa. In the extreme example of 6 January 1665, when the planets subtended an angle of 45° to the Earth, it is estimated that the winter half of the year was increased by almost two days with a corresponding shortening of the summer half of the year. This is a potentially significant physical shift and may in part explain why this synod marked the onset of the coldest period of the Little Ice Age in the northern hemisphere.

The other possibility of planetary motions causing the observed cyclic behaviour of sunspot numbers has been the subject of considerable debate.[28] The periods of the orbits of Jupiter, Uranus and Neptune coincide roughly with the 11-, 90- and 180-year cycles in the sunspot records, and this has led to a planetary theory of sunspots. Of particular interest are the effects of the planets on the motion of the Sun around the centre of mass of the solar system. Calculations show that this complicated motion is dominated by the orbits of Jupiter and Saturn, and, in particular, the time taken for Jupiter to lap Saturn – 19.9 years. But over the last 1200 years the period of the Sun's motion has varied between 15 and 26 years. The other important cycle is 177.9 years, which is the product of the near coincidence of 15 Jupiter orbits and 6 Saturn orbits. The consequence of the Sun's motion is to affect its oblateness, diameter and rate of spin, all of which could influence the sunspot mechanism. So we cannot examine solar variability and tidal forces in isolation.

6.3 Physical links between solar and tidal variations and the weather

Having identified the most important aspects of the periodic variation of both solar activity and the tidal forces acting on the Earth, we must now look more closely at how they could influence the weather. In particular, the

[28] See for example Okal & Anderson (1975) and also Landscheit in Rampino *et al.* (1987), pp. 421–45.

periods of 11, 18.6, 22, 90 and around 200 years in these external influences appear to be reflected in observed fluctuations in the weather. So we need to find a reasonable explanation of how these external changes alter the weather. In so doing we must address the fundamental problem that these changes are very small and that without some amplification process it is hard to see how they produce significant weather effects. In short, how does the tiny tail wag the large dog? At the same time we must address an additional complication. This is the fact that in the many examples of statistical correlations between aspects of solar behaviour and features of the weather there appears to be a delay between the one and the other: so, not only must we explain how the tail can wag the dog but also why in some instances it takes a long time for the dog to react to the stimulus of the wagging tail.

Starting with the observed total solar irradiance (TSI), the task is first to identify just how much the TSI has varied in recent centuries and then to propose physical mechanisms that could translate these small changes into significant climatic shifts.[29] Although changes observed since 1980 are not sufficient to explain the significant correlation between longer-term solar activity and global temperatures, the estimates of the larger variations in TSI since the Maunder Minimum (see Section 6.1) go some way towards substantiating the hypothesis that this aspect of solar variability has played a significant part in climatic change over the last few centuries. Beyond this, we must turn to other climatic mechanisms to find a physical explanation of the correlation between solar variability and global climate change.[30]

The most obvious approach is to focus on the fact that much of the variability in TSI is concentrated in the UV (see Section 6.1). This affects the chemistry of the upper atmosphere, and, in particular, leads to changes in the production of ozone. The evidence of how ozone concentrations in the stratosphere shift with UV variations is complicated. Satellite data suggest the largest changes occur in the upper stratosphere, with zero, or even slightly negative, changes in the middle stratosphere and significant positive changes in the lower stratosphere. However, as the data are only available over about one and a half solar cycles, have large uncertainties, especially in the lower stratosphere, and may not properly have accounted for the effects of volcanic aerosol from the Mount Pinatubo eruption in 1991. Thus, the true nature of solar-induced changes in stratospheric ozone remains uncertain.

---

[29] Haigh (2000) and Soon, Posmentier & Baliunas (2000a).

[30] Much of this section draws on Chapter 6, Section 6.11 of the IPCC Third Assessment Review (2001), but where specific papers are used to make more detailed comments, they are identified in separate footnotes.

The amount of ozone in the stratosphere affects how much UV radiation in the wavelength range 280 to 310 nm is absorbed in the stratosphere. In addition, because ozone absorbs strongly in the mid-infrared region where the Earth reradiates to space, it can alter the radiative balance of both the stratosphere and lower levels of the atmosphere. Any decrease in ozone levels in the stratosphere will lead to a direct increase in downward short-wave irradiance at the tropopause, and vice versa. At the same time, changes in stratospheric ozone will alter the thermal structure of these atmospheric levels and hence the amount of downward infrared flux into the troposphere.

Work with computer models by Joanna Haigh, at Imperial College, London, England, and Drew Shindell and colleagues at the NASA Goddard Institute for Space Studies, New York, which include realistic changes in UV and ozone, shows a significant effect on simulated climate.[31] The predicted changes consist of a warming in the stratosphere (except in winter at high latitudes) and a vertical banding structure in the troposphere due to shifts in the positions of the sub-tropical jet streams. This could lead to dipole anomalies in the zonal wind structure that propagate down over the winter period, and modulate tropospheric circulation patterns such as the Arctic Oscillation (see Section 5.7). Solar cycle variability may therefore play a significant role in regional surface temperatures, even though its influence on the global mean surface temperature is small. So it is possible, by both radiative and dynamical means, that variations in solar UV output can produce disproportionate changes in the atmosphere. The fact that ozone and other chemical species of the stratosphere vary in phase with the QBO[32] may also be the key to the QBO–solar link in the upper atmosphere.

Solar magnetic field variations could lead to more complicated mechanisms. Given the amount of evidence that the 22-year magnetic (Hale) cycle is present in weather statistics, these proposals need careful consideration. Variations in the solar magnetic field in the vicinity of the Earth alter the flux cosmic rays that funnelled down into the atmosphere. This is a complicated process as cosmic rays come in a wide variety of forms. They consist mainly of protons, with smaller amounts of helium and heavier nuclei. Cosmic rays of low energy have their origin in the Sun and are absorbed high in the atmosphere. It is these particles that are the origin of the aurora during periods of high solar activity active. Galactic cosmic rays (GCRs) are of much higher energy and have an appreciable impact on the

[31] Haigh (1999), and Shindell *et al.* (1999).
[32] See section 5 of Baldwin *et al.* (2001) for an extensive review of the satellite evidence of these changes.

troposphere. When the Sun is more active, GCRs are less able to reach the Earth and so their impact on the lower atmosphere is inversely related to solar activity.

Cosmic rays produce various chemical species such as NO, OH and NO_3 that can catalyse chemical reactions. This leads to changes in the atmospheric concentrations of radiatively active molecules such as ozone (O_3), nitrogen dioxide (NO_2), nitrous oxide (N_2O) and methane (CH_4). These species are most likely to be seen in the stratosphere where their impact will be similar, but less significant, to the changes caused by solar UV variations. In addition, the formation of ions will affect the behaviour of aerosols and cirrus clouds that have a direct radiative impact and also alter the amount of water vapour throughout the atmosphere. These changes could lead to shifts in the radiative balance of both the stratosphere and troposphere and so produce long-term fluctuations in the temperature.

The most controversial aspect of these changes is the question of whether GCRs alter cloud cover at lower levels. A study by Henrik Svensmark of the Danish Space Research Institute (DSRI), Copenhagen, Denmark, and Eigil Friis-Christensen, Director of the Danish Space Research Institute (DSRI),[33] demonstrated a high degree of correlation between total cloud cover, from satellite studies, and cosmic ray flux between 1984 and 1991. Their analysis of monthly mean data of total cloud using data over the oceans between 60° S and 60° N from geostationary satellites found an increase in cloudiness of 3% to 4% from solar maximum to minimum. They proposed that increased GCR flux causes total cloud amounts to rise and this cools the climate. Other satellite data appeared to support this conclusion, and, in addition, total cloud varies more closely with GCRs than with the 10.7 cm solar activity index over the past solar cycle. More recently, however, the correlation has been far less impressive. On longer timescales Svensmark also demonstrated that northern hemisphere surface temperatures between 1937 and 1994 follow variations in cosmic ray flux and solar cycle length more closely than total irradiance or sunspot number.

The interannual variations in cloudiness are, however, difficult to distinguish from parallel changes caused by warm and cold ENSO events. Also, the correlation with cosmic ray flux tends to be reduced if high latitude data are included. This would not be expected if cosmic rays were directly inducing increases in cloudiness, as cosmic ray flux is greatest at high latitudes. Moreover, a mechanism whereby cosmic rays resulted in greater cloud cover would be most likely to affect high cloud as ionisation is greatest at these altitudes. But, if high cloud does respond to cosmic rays

[33] Svensmark & Friis-Christensen (1997).

it is not clear that this would cause global cooling, as for thin high cloud the long-wave warming effects dominate the short-wave cooling effect. If there is an answer to these objections, it may lie in unscrambling the range of impacts solar particles and GCRs will have on the atmosphere at different altitudes and latitudes. Before going down this route, however, we need to weave in a few more examples of GCR–weather correlations.

One example of the longer-term changes in cosmic ray flux can be seen in the ^{10}Be and ^{14}C isotope records, which appear to correlate well with observed climatic shifts in the North Atlantic.[34] These changes suggest that, during the Holocene at least, the 1450-year cyle appears to have been linked to variations in solar irradiance. They also show variations that are consistent with estimates of the increase in TSI since the seventeenth century. These changes may also tie in with what was said in Section 6.1 about the Sun's magnetic field at the Earth (the aa geomagnetic index),[35] which has been measured since 1868. An intriguing feature of changes in this index is that calculation of the correlation between the annual value of this index and the value of annual northern hemisphere temperature anomaly (NHTA) reveals an unexpected behaviour. This is that the square of the correlation coefficient between the annual value of the aa index and the subsequent value of the NHT peaks for a lag of 6 years with a value of 0.466 (Fig. 6.10). This correlation declines to values of 0.26 or less for lags of less than 3 years and greater than 10 years. No equivalent variation is found in the case of sunspot numbers. This suggests that some 20% of all variance in the NHTA record, and much of the variability with periodicities in the range 3 to 10 years, could be associated with variations in the aa index. What is far from clear is what combination of lags and leads in the climate system could result in such a delayed reaction to solar magnetic influences. But as was apparent in Chapter 5, there is no shortage of delayed teleconnections in the global climate system.

Any lagged relationship between the solar magnetic field and global temperature anomalies will also have to be reconciled with another interesting and apparently highly significant correlation. This is between the solar coronal hole area (see Section 6.1) and measurements of the terrestrial lower tropospheric air temperature obtained by microwave sounding units (MSU) on orbiting weather satellites,[36] which is an unlagged response, suggesting that in respect to the effects of coronal holes the climatic reaction is rapid. Using records of coronal holes and MSU measurements of global

---

[34] Bond *et al.* (2001).

[35] Lockwood & Stamper (1999) and Lockwood, Stamper & Wild (1999).

[36] Soon *et al.* (2000b).

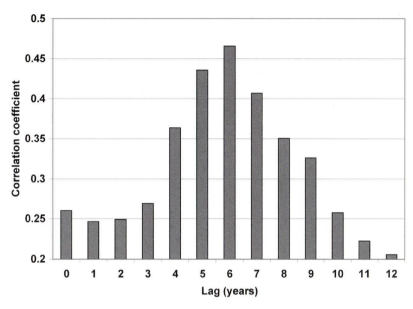

Fig. 6.10. The square of the correlation coefficient between the annual values of northern hemisphere temperature anomaly and the annual values of the aa geomagnetic index between 1880 and 2000 as a function of the lag (in years) between the aa index values and the subsequent temperature anomalies, showing how this relationship has a marked peak for a lag of 6 years.

temperature from 1979 to mid-1998, this work provides evidence of not only a close correlation between the solar wind plus its associated magnetic field and global tropospheric temperature, but also interesting variations in the level of correlation, with the strongest signal being seen in the northern and southern mid latitudes, and a less strong signal in the tropics. These results can be interpreted as being consistent with the hypothesis that global temperature changes are affected by solar particles and GCRs.

Whatever the explanation for this polyglot set of observations, there remains the basic question of how solar magnetic fluctuations can influence the weather by, say, altering cloudiness. To address this issue, we need to consider how the Sun's magnetic field interacts with the Earth's magnetic field and how this can lead to physical effects that could affect the weather. Ralph Markson at the Massachusetts Institute of Technology[37] developed an influential proposal, which sought to link solar magnetic variability to changes in the weather. He was particularly interested in Walter Orr Roberts' proposals that certain features of the weather respond to the passage of sector boundaries of the Sun's magnetic field within a day or so

[37] Markson (1978).

of their crossing the Earth (see Section 6.1). He argued that not only are the changes in energy flux too small to produce changes in weather, but also the response to short-term fluctuations in solar activity is too rapid to be the consequence of changes in energy input. This suggested that an electromagnetic explanation was needed. In this context, Markson proposed that the modulation of cosmic rays by the Sun leads to changes in the Earth's electric field and hence thunderstorm activity or frequency would be altered.

This mechanism had three attractions. First, it required no significant change in solar energy to alter the state of the Earth's magnetic field and stratospheric conductivity, while offering the possibility of releasing and redistributing large amounts of energy already present in the troposphere. Second, it did not require strong links between the upper and lower atmosphere as the electric field variations encompass the whole atmosphere from the ionosphere to the Earth's surface. Third, the response of the electric field to changes in the magnetic field is almost instantaneous and so offers an explanation of how the weather can respond within a day or so to changes in solar activity.

Markson postulated that worldwide, thunderstorms play a central role in maintaining a global electric circuit. What is not clear is how the changing of the conductivity of the upper atmospheres can alter the incidence of thunderstorm activity. While it is possible that greater stratospheric ionisation could lead to increased thunderstorm electrification either locally or as a consequence of global changes in the Earth's electrification field, it is not clear how this will alter thunderstorm development. Moreover, the changes in the Sun's magnetic field and related changes in the number of energetic particles emitted by the Sun will have complicated effects on the Earth's atmosphere. For while at high latitudes the effects of the flux of solar protons, which is directly related to solar activity, will predominate, at low latitudes the magnetic field variations will modulate GCRs and produce an effect that is out of phase with solar activity. This may explain why there is evidence of a high correlation between the sunspot cycle and high-latitude thunderstorm activity, whereas at low latitude it is either non-existent or negative.

Although this theory was not pursued at the time, it has stimulated work that has focused on the possibility that charged particles were more efficient than uncharged ones in acting as cloud condensation nuclei (CCN). In particular, Brian Tinsley at the University of Texas developed a detailed mechanism for a link between cosmic rays and cloudiness.[38] This was

[38] Tinsley (1996).

based on the premise that aerosols ionised by GCRs are more effective as ice-forming nuclei and cause freezing of supercooled water in clouds. In clouds that are likely to cause precipitation the latent heat thus released then causes enhanced convection that promotes cyclonic development and hence increased storminess.

The IPCC Third Assessment Report[39] concluded that these proposals required more research to establish whether or not the link between cosmic rays and cloud formation could be of sufficient magnitude to result in the claimed effects. Although there was some laboratory evidence to suggest that charging increases ice nucleation efficiency, at the time there was no observational evidence of this process taking place in the atmosphere. Furthermore, only a small proportion of aerosol particles is capable of acting as ice nuclei, depending on chemical composition or shape. There are also laboratory studies[40] which indicate the existence of 'electrofreezing' – the electrical enhancement of the phase transition between supercooled water and ice – but again no evidence in the real atmosphere. All this may have changed with results obtained by Frank Arnold's group at the Atmospheric Physics Division, at the Max-Planck-Institut of Nuclear Physics, Heidelberg, Germany, that has provided the first observational evidence of cosmic ray-induced aerosol formation in the upper troposphere.[41]

In the meantime, Brian Tinsley has continued to refine these cosmic ray-induced aerosol formation hypotheses. Together with Fangqun Yu, at the State University of New York, at Albany,[42] he argues that, while ions cannot act as efficient condensation sites for water vapour, they can act as condensation sites for sulphuric acid vapour and then with water vapour grow into aerosols, which, under certain conditions, grow to sufficient sizes to act as CCN. This process, which is defined as 'ion mediated nucleation' (IMN), appears to underlie the formation of ion clusters that have been observed by Frank Arnold's group. These clusters are formed by a number of water vapour molecules plus some organic molecules (mostly acetone) and sometimes sulphuric acid molecules. The latter are formed in the upper troposphere by photochemical conversion of sulphur dioxide, which in the northern hemisphere stems mainly from fossil fuel combustion.

In addition, Tinsley and Yu propose that aerosols are involved in a process of 'electroscavenging' – a possible electrofreezing mechanism whereby electrical charging of water droplets and aerosols increases their collision efficiency. In theory, a higher the level of charge carried by an aerosol leads

---

[39] IPCC (2001). See Chapter 6, p384.   [40] Abbas & Latham (1969).
[41] Eichkorn *et al.* (2002).   [42] Tinsley & Yu (2002).

to greater the collection efficiency.[43] This process may also enhance the capture of those rare aerosols suitable as ice nuclei. The level of charge on aerosols depends on the net space charge in clouds, which is determined by the ionospheric–earth current density. The space charge that builds up at the top and bottom of clouds is electrostatic charge density due to a difference between the concentrations of positive and negative ions. This charge density is transferred to charges on aerosol particles (both ice forming nuclei and CCN), and this affects the rate at which they are mopped up by collisions with water droplets and ice crystals. Both IMN and electroscavenging depend on the presence of atmospheric ions that are generated by GCR flux. In addition, the ionospheric–earth current depends not only on the GCR flux but also on the strength of the solar wind.

Unlike Markson' hypothesis, which focused on changes in thunderstorms, this approach is of more relevance to weakly electrified, non-thunderstorm clouds, such as marine stratocumulus, or nimbostratus, of much greater geographic and temporal extent, which are affected by changes in the ionosphere–earth current density and cosmic ray flux. By changing precipitation rates or radiative balance, the changes in the clouds then affect atmospheric dynamics and temperature. But, until we have confirmation that changes in atmospheric electricity lead to changes in patterns of cloudiness, we have to treat all these theories with caution. What is important, however, is that the possibility of electrical modulation of the atmosphere being amplified to produce more substantial changes in the climate is an elegant way of getting round the basic objection that solar cycles contain insufficient energy to produce the observed climatic fluctuations often attributed to them.

There are two additional features of magnetic field changes that need to be considered here. The first is that they will not be symmetric about the Earth's axis and this will affect how cosmic rays modify the upper atmosphere (Fig. 6.11). Because the geomagnetic poles are not coincident with the geographic poles, the perturbations of the magnetic field will be off axis. The fact that the circulation pattern in the northern hemisphere shows a similar off-axis form (see Section 5.1) may also be of relevance. The observed consequences of the link between the QBO and the 11-year sunspot cycle (Section 3.10) are reflected in both the circulation over North America centred on the geomagnetic pole and the latitude of the winter storm track across the North Atlantic. So, alongside the various cloud-modification theories, we may be looking at signals of solar magnetic field changes in mid-latitude circulation patterns. In this context, analysis of the correlation

---

[43] Tripathi & Harrison (2002).

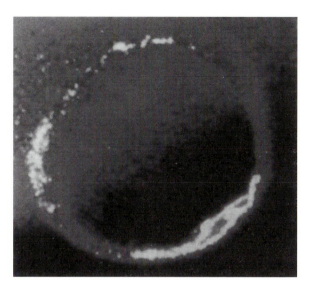

Fig. 6.11. A satellite ultraviolet image of the auroral oval. This shows that the auroral activity encircles the geomagnetic poles, but that the maximum activity is centred near local midnight and local noon. (From Lundin, Eliasson & Murphree, 1991.)

between cosmic ray flux and precipitation in the southern hemisphere[44] shows the most significant figures at 65–75° geomagnetic latitudes where the GCR flux varies by 15–20%. Similarly, as discussed above, the correlation between solar coronal hole area and tropospheric temperature[45] shows the highest values in northern and southern mid latitudes.

The second consequence of magnetic field changes is the long-term decline in the fair-weather potential gradient in the twentieth century.[46] This measure of the strength of the global electrical circuit appears to be linked to the parallel decline in cosmic ray flux over the same period. So to the extent that this is a measure of how solar activity could influence global cloudiness it provides additional insight into the processes that may be leading to longer periodic fluctuations in the Earth's climate.

To the extent that thunderstorms play a part in all these processes the key to monitoring their activity may lie in an unexpected area. This involves an intriguing property of the global atmosphere. The thin shell formed by the atmosphere between the Earth's surface and the ionosphere, at an altitude of some 80 km, forms a waveguide for very-low-frequency electromagnetic radiation. In particular, a frequency of around 8 Hz has a wavelength of 40 000 km (the circumference of the Earth) can ring round the world with virtually no loss. Known as 'Schumann resonances' after the

[44] Kniveton & Todd (2001).   [45] Soon *et al.* (2000b).   [46] Harrison (2002).

German scientist who first proposed this phenomenon in 1952, this signal can be measured by electromagnetic detectors. The dominant source of such radiation is lightning in thunderstorms. So, at any time, the size of the 8-Hz signal is a direct measure of worldwide thunderstorm activity.[47]

The potential importance of this phenomenon is that not only is it a measure of the global level of thunderstorm activity but also this activity appears to be peculiarly sensitive to the mean global temperature – a 1°C rise will led to a 10% rise in the number of thunderstorms. So the measurement of Schumann resonances offers the prospect of both checking whether the level of global thunderstorm activity is affected by solar activity and also providing a different means of monitoring global warming.

Returning to the issue of tidal forces, the obvious link between these forces and the weather is the direct movement of the atmosphere, the oceans and even the Earth's crust. The nature of these links varies in complexity. Tidal effects in the atmosphere are relatively predictable and measurable but tiny compared with normal atmospheric fluctuations (see Section 7.4). In the oceans the broad effects can be calculated, but estimating changes in the major currents is much more difficult. In particular, recent satellite altimetry studies have come up with some interesting results that may shed new light on the climatic implications of the dissipation of tidal energy.[48] Prior to this work it was widely assumed that this energy, representing the effect of the Moon receding from the Earth at a rate of some four centimetres a year, was dissipated in the shallow waters of the continental shelves around the world. It is now reckoned that about half this energy is fed into deeper water, where it exerts a significant effect on the strength of the major ocean currents and hence the transport of energy from the tropics to polar regions (see Section 7.4). So we now need to look more closely at evidence of whether, say, the 18.61-year cycle can be detected in the strength of the principal ocean currents, because, if so, this would be a potentially important way for lunar tides to affect the climate.

When it comes to movements of the Earth's crust, the problems are compounded by possible links with solar activity. The direct influence of the tides could influence the release of tectonic energy in the form of volcanism.[49] Since there is considerable evidence that major volcanic eruptions have triggered periods of climatic cooling, this would enable small extraterrestrial effects to be amplified to produce more significant climatic fluctuations. In addition, there is evidence that intense bursts of solar activity interact with the Earth's magnetic field to produce measurable changes in the length of the day. Such sudden tiny changes in the rate of rotation of

[47] Williams (1992) and (1994).   [48] Wunsch (2000).   [49] Roosen *et al.* (1976).

the Earth could also trigger volcanic activity. It should be noted that, while there is no evidence that their occurrence was in any way related to this effect, the three climatically important volcanoes in the second half of the twentieth century (Agung in 1963, El Chichón in 1982 and Pinatubo in 1991) were spaced in such a way as to have a confusing impact on the interpretation of any solar or tidal effects. So, this is yet one more complication that must be considered when seeking to identify the cause of oscillations in the climate.

The most convincing evidence of a periodicity in volcanic activity has emerged from recently published analysis of ice cores from high latitudes in both hemispheres.[50] The work identified the 61 largest tropical volcanic eruptions in the last 600 years using the qualification that all of them produced an identifiable sulphate signal in at least one ice core in each hemisphere. This approach is likely to identify all the eruptions which had a global impact on the climate during this period, although it is possible that effectively simultaneous eruptions in both hemispheres could also be part of this analysis, but the chances of this having happened is small. What the study shows is a remarkably strong 76-year cycle that appears to demonstrate a three-fold difference in the incidence of major tropical eruptions between the peaks and the troughs of the cycle (rising and falling between one eruption every 5 years down to on every 15 years). There is, howerver, one difficulty with this periodicity as far as tidal effects are concerned. There is no obvious reason why the tides should be the cause. Indeed, the nearest possibility is the 90-year solar cycle, and linking that with the release of tectonic activity poses a much greater challenge.

6.4 Orbital variations

The Earth's orbit around the Sun is also influenced by the gravitational interactions with the Moon and the other planets on much longer timescales. The resulting perturbations give rise to a set of cyclical variations in orbital eccentricity, obliquity and precession, the principal periods of which occur at around 413, 100, 41, 19 and 23 kyr. These variations are climatically important as they affect the seasonal and latitudinal distribution of solar radiation. As noted in Section 4.8, the theory of these climatic effects is usually attributed to the Serbian geophysicist Milutin Milankovitch, who transformed the earlier semi-quantitative work of James Croll into a mathematical framework of an astronomical theory of climate. It is a theory that

[50] Ammann & Naveau (2003).

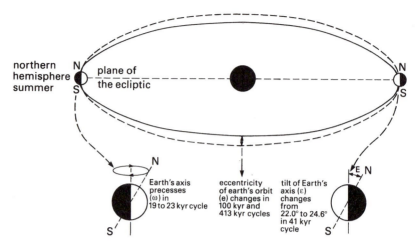

Fig. 6.12. The changes in the precession, tilt and shape of the Earth's orbit which are the underlying cause of longer-term variations in the climate.

has been refined since the 1960s to provide an explanation for the observed variations in the climate over the last million years. The description here and the climatic consequences outlined in Chapter 7 draw heavily on the work of John Imbrie and John Z. Imbrie of Brown University, Rhode Island, and Harvard University, respectively, in the early 1980s.[51]

If the output of the Sun is assumed to be constant, the amount of solar radiation striking the top of the atmosphere at any given latitude and season is fixed by three elements. First, there is the eccentricity (e) of the Earth's orbit. Second is the tilt of the Earth's axis to the plane of its orbit – the obliquity of the ecliptic (ε). Third is the longitude of the perihelion of the orbit (ω) with respect to the moving vernal point (Fig. 6.12). Integrated over all latitudes and over an entire year, the energy flux depends only on e. However, the geographic and seasonal pattern of irradiation depends on ε and $e \sin \omega$. The latter is the parameter that describes how the precession of the equinoxes affects the seasonal configuration of the Earth–Sun distances. For the purposes of computation the value of $e \sin \omega$ at AD 1950 is subtracted from the value at any other time to give the precession index $\delta\,(e \sin\omega)$. This is approximately equal to the deviation from the value of the Earth–Sun distance in June 1950, expressed as a fraction of the invariant semi-major axis of the Earth's orbit.

Each of these elements is a quasi-periodic function of time (Fig. 6.13). Although the curves have a large number of harmonic components, their power spectra are dominated by a small number of features. The most

[51] Imbrie & Imbrie (1980).

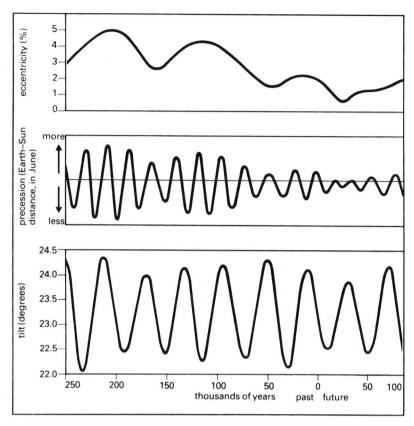

Fig. 6.13. Calculated changes in the Earth's eccentricity, precession and tilt. These changes reflect the fact that the Earth's orbit is affected by variations in the gravitational field due to planetary motion. (From Smith, 1982.)

important term in the eccentricity spectrum has a period of 413 kyr. Eight of the next 12 most significant terms lie in the range from 95 to 136 kyr. In low-resolution spectra these terms contribute to a broad peak that is loosely referred to as the 100-kyr eccentricity cycle. In contrast, as can be seen in Fig. 6.13, the variation of the obliquity is a much simpler function and its spectrum is dominated by components with periods near 41 kyr. The precession index is an intermediate case with its main components being periods near 19 to 23 kyr. In the low-resolution spectra this can be seen as a single peak near 22 kyr.

Calculations of past and future orbits provide figures for the variation of the orbital elements. The present value of e is 0.017. Over the past million years it has ranged from 0.001 to 0.054. Over the same interval ε, which is now 23.4°, has ranged from 22.0 to 24.5°, and the precession index (defined as zero at AD 1950) has ranged from –6.9% to 3.7%. Because these variations alter the seasonal and latitudinal input of solar radiation to the top of the atmosphere they will bring about some form of climatic change. Most obviously the changes in the obliquity will have seasonal effects. If the obliquity was reduced to zero, the seasonal cycle would vanish and the pole-to-equator contrasts would sharpen. So low values of obliquity should correlate with colder periods at high latitudes, which is indeed the case. The eccentricity also exerts a seasonal influence. If e was zero and the Earth had a circular orbit about the Sun, there would be no seasonal effect from this source. The precession of the orbit means that if the summer solstice were shifted towards the perihelion and away from its present position relatively far from the Sun, summers in the northern hemisphere would become warmer, and winters colder than they are today.

As Milankovitch recognised, the key to how these variations can precipitate ice ages is in the intensity of radiation received at high latitudes during the summer. This is critical to the growth and decay of ice sheets. For whereas at all latitudes over the last 600 kyr the irradiation intensity during the summer and winter half of the year has varied by more than 5%, at 65° N the variations have exceeded 9%. In spite of the scale of these variations, which were calculated by Milankovitch in 1930, attempts to translate these figures into climatic models which explained the ice ages did not make rapid progress. During the 1960s and early 1970s, it was concluded that the climatic response to orbital changes was too small to account for the succession of ice ages. But as a result of new perceptions about the more frequent cyclic nature of past ice ages (see Section 4.8), the Milankovitch theory was re-evaluated. Experiments with a new generation of models (see Section 7.5) now suggest that these orbital variations are sufficient to explain past major changes in the size of the northern hemisphere ice sheets.

Chapter 7

Autovariance and other explanations

Hereafter, when they come to model heaven
And calculate the stars, how they will wield
The mighty frame, how build, unbuild, contrive
To save appearances, how gird the sphere
With centric and eccentric scribbled o'er
Cycle and epicycle, orb in orb.

Milton (*Paradise Lost*)

Having considered the evidence of cycles and the possible mechanisms for the occurrence of natural variability of the climate and the sources of external periodic behaviour, the final exercise is to identify how a physical model or other explanation can link these various observations. But this objective requires first a review of just what features of the climate should be covered by any models. This in itself is not easy. As has now become evident, neither the evidence of cycles nor the possible causes of such behaviour present a simple picture. So it is important to clarify this objective from the outset.

Clearly, the evidence of cycles is not unequivocal. What can be said is that there is a strong need to find an explanation of the QBO. But, as has become apparent, this is not a sharply defined periodicity; moreover, while the stratospheric phenomenon is well defined, its equivalent in the troposphere is an altogether more shadowy creature. So, in the stratosphere we are looking for a physical explanation of a strikingly regular and substantial feature of the equatorial regions. By way of contrast, in the troposphere we are trying to identify whether a general presence in many meteorological records of a significant oscillation of around 2.1 to 2.5 years, which may or may not be related to the QBO in the stratosphere, can be explained in a coherent manner. The absence of a distinct frequency makes the search more difficult as it tends to rule out a single predictable mechanism. This challenge may extend to the ill-defined periodicities that frequently appear in the range 3 to 4 years, around 5 years and sometimes at 7 to 8 years.

In moving to longer periodicities a different problem arises. Here, the challenge becomes one of discrimination. We have to decide whether the apparently well-defined periodicities are the product of simple climatic variability, or whether there is a credible physical link between them and extraterrestrial influences such as solar variability and tidal effects. So it is necessary to show not only that the cycles in the weather records closely match these external variations, but also that there is a satisfactory physical link between the two effects. It is not enough to rely solely on the coincidence, which, given the somewhat limited evidence of the required periodicities, is bound to be of questionable value. If, however, it is possible to postulate a realistic and physically plausible link between the two processes then the significance of the observed weather cycles is much greater. This is of the greatest importance when providing an explanation of the periodicities in the range 10 to 12 years and around 20 years.

Underlying these specific efforts to explain certain periodicities is a more general requirement. With the exception of the climatic variations associated with the ice ages, only a small part of the variability is found in discrete features. For the rest, the majority of the variance is in a broad-band spectral continuum. So it is important that the explanations tendered here address not only the significant sharp spectral features but also the scale and frequency distribution of the 'background noise'.

7.1   Non-linearity

Before considering any specific models or explanations, we must address the matter that has lurked beneath the surface of all aspects of weather cycles. This is the question of non-linearity. This issue is fundamental to the understanding of cyclic and quasi-cyclic behaviour in the weather. It has already been touched upon in a number of places in the book. Now we must look more closely at its implications for explaining the various phenomena that have been described so far.

If the relationship between two or more variables is strictly proportional, the relationship is said to be 'linear'. This means that when one variable is altered by a given amount other related variables will respond by a simple multiple of this change. In many instances this is not the case, and the relationship between variables is 'non-linear', in other words the description of the relationship on a graph would be a curve. It is possible to calculate the behaviour of systems governed by non-linear relationships where we are considering only two variables. More generally, for complicated systems governed by a large number of non-linear equations, where the relationship between parameters lacks direct proportionality, the equations are insoluble. Under certain limited circumstances it proves to be an acceptable approximation to assume that linear relationships can apply for small perturbations. As a consequence, some of the successes of physics rely on the fact that this simplification can produce excellent descriptions of simple systems. In practice, however, most of the real world is both complicated and runs on non-linear lines. As such it is inherently unpredictable, although not necessarily chaotic (see Section 1.2). Nowhere is this truer than with the weather.

The consequences of the non-linearity of the climate change are profound, as will be seen in Chapter 8. But although non-linearity places major limitations on our understanding of the weather, it does not put a total restriction on what we can say about cycles. Because of work done principally in the field of electrical engineering, quite a lot is known about what happens to cycles in non-linear systems.[1] The guidelines, while saying nothing about the predictability of the climate, do show what sort of phenomena might occur in non-linear systems, which tend to have natural oscillations and are also subject to external forced oscillations.

The most obvious effect of non-linear systems is harmonic generation. Non-linear systems produce higher harmonics when forced to oscillate at a given frequency, and when the system is excited at two or more frequencies they produce sums and differences of these frequencies. Simple harmonic generation produces multiples of the fundamental frequency. The amplitude of the higher harmonics will depend on the non-linearity of the system, but will, in general, decrease rapidly with increasing frequency. The sum and difference effects (often termed 'frequency demultiplication') are best described in terms of two frequencies (v_1 and v_2). Non-linear systems acted upon by two such periodic inputs will generate not only harmonics of these frequencies but also a whole range of combinations given by the general expression $mv_1 \pm nv_2$ where m and n are integers. The important feature is the 'subharmonics' such as $v_1 - v_2$, $2v_1 - v_2$, $v_1 - 2v_2$, and so on, which can produce low-frequency oscillations.

The implications of these processes are best seen by considering how they might operate on sunspot and lunar cycles. In the case of sunspots, the frequencies of the 11-year and 22-year cycles are 0.091 cpa and 0.045 cpa respectively. The interesting combinations are the second and third harmonics of the 11-year cycle with periods around 5.5 and 3.7 years, and the sum of the 11- and 22-year cycles, which would produce a frequency around 0.136 cpa or a periodicity of 7.4 years. The sums and differences of the

[1] Tsien (1954).

18.6-year lunar cycle and the two sunspot cycles could produce periodicities of 6.9, 10.1, 26.9 and 120.3 years. But this would require the climate to respond to the primary frequencies, which, as earlier chapters have demonstrated, is less than abundantly clear. So the existence of the sum and difference frequencies is a matter of even greater conjecture.

Another interesting frequency response is known as 'entrainment'. If a system that has a natural self-excitation frequency v_1 is subjected to an input of a slightly different frequency v_2, the system may not behave in the way described above. Instead of both v_1 and v_2 and the difference frequency $v_1 - v_2$ being present, the whole system may oscillate at v_2 with the original self-excitation oscillation effectively entrained by the imposed frequency. The range of frequencies over which this phenomenon can occur depends on the properties of the system and is known as the 'zone of synchronisation'. A related but more unlikely effect is that in some non-linear systems it is possible to start or stop an oscillation by starting up an entirely different frequency. This excitation or quenching is an entirely arbitrary consequence of the system, and is usually termed 'asynchronous' to reflect its unpredictable nature.

One final form of behaviour worth mentioning is the differing response of systems to self-excitation. Some require only small oscillations from equilibrium to build up. This is known as 'soft excitation'. Other systems require much greater perturbations before they break into oscillation. This 'hard excitation' then appears with a sudden jump. Conversely, it will exhibit hysteresis in that, as the oscillation decays, the system will continue to oscillate at a lower amplitude than the initial threshold needed originally to get it going. This variable or erratic response to both self-excitation and forced oscillations is yet another indication of the unpredictable nature of such systems. Such behaviour is also reflected in another aspect of the reaction of a system to forced oscillations. At the nodal point, when the response of the system is zero, the phase of the response can shift by π radians. This is more probable when the system is being driven by a combination of frequencies that slowly move in and out of phase. For example, the 8.85-year and 18.6-year lunar cycles can combine to produce a 165-year overtone. So it is feasible that at a certain point in the long periodicity the response of the climate to the 18.6-year can switch completely. As noted in Section 4.1, there is clear evidence that the behaviour of the drought index in the United States could be explained by this phenomenon.

These basic examples of what might be regarded as the predictable behaviour of non-linear systems were originally studied as part of the development of electrical and mechanical control equipment. They show how oddly even relatively simple systems can behave when subjected to periodic signals. As will be seen in Chapter 8, the behaviour of non-linear systems is even less predictable than these physical examples suggest. But for the moment these models are helpful as their implications for weather cycles are obvious. The global climate is both non-linear and immensely more complicated than the electronic circuits studied in the development of cybernetics. So it would hardly be surprising to discover that the weather is capable of exhibiting all these forms of behaviour. As a consequence, we must expect any periodicity to combine with others to form a gamut of oscillations. Moreover, they may come and go in odd ways as and when the balance of climatic conditions may temporarily be suited to the establishment of certain periodicities. Against this fluctuating background we can now consider the various possible reasons for the existence of weather cycles.

7.2 Natural atmospheric variability

So far there has been a general presumption that the natural variability of the atmosphere is too short term to explain the observed fluctuations with periods of several years. But it has been noted that stable atmospheric patterns ('blocking') can persist for several months. Indeed, the most famous example of such a pattern exists not on Earth but on Jupiter. The great Red Spot has been a perennial feature of the planet's circulation since it was first observed in the seventeenth century. Moreover, laboratory fluid-dynamical experiments have produced similar stable patterns in rotating differentially heated systems.[2] So, in principle, it is possible that fluctuations in the Earth's atmospheric patterns alone could explain some of the long-term periodicities or quasi-periodicities observed in weather records.

The possibility that the variability of atmospheric flow at low frequencies may be due to the inherent, internally generated unsteadiness of that flow rather than external influences in the form of varying boundary forcing (e.g. sea surface temperatures, SSTs) or changing external influences (e.g. solar or tidal) has been explored by Ian James and Paul James of the Department of Meteorology at University of Reading, UK.[3] This study noted that the relative distribution of internal and external sources of variability was uncertain. Arguments based on linear theories of instability and on the known limits of deterministic predictability suggest that the internal variability of the atmosphere dominates for periods of less than 10 to

[2] Sommeria, Meyers & Swinney (1988) and Marcus *et al.* (1990).
[3] James & James (1989).

20 days. The study then went on to show that internally generated variations of much lower frequencies can dominate the spectrum of planetary-scale structures in a non-linear atmospheric model. This implies that a degree of regularity underlies the apparently random fluctuations in the circulation.

Although the model was relatively crude, in that it had a horizontal resolution of 700 km, it handled the non-linear dynamics in a reasonably sophisticated manner, explicitly representing the large-scale circulation systems of both the tropics and mid-latitudes. Simple linear terms were adopted for the physical treatment of heating and friction due to the Earth's rotation to enable extremely long integrations to be conducted at modest cost. This simple model ignored moisture and made the lower surface uniform without mountains or heat sources and sinks. In addition, the seasonal cycle was represented by a sinusoidal variation of the pole-to-pole radiative equilibrium temperature difference. In spite of this basic approach, the model is a powerful technique for examining the spectrum of atmospheric fluctuations.

The fascinating result of this modelling work is that it shows that atmospheric circulation varies strongly for periods longer than a year with the strongest response being around 10 years. This general result was obtained from a number of experiments. Various measures of the large-scale atmospheric flow were studied by collecting the average data for every 10 (model) days and then calculating the Fourier transform of the data for 96 years once the model had settled down and eliminated any initial transient start-up effects. An example of the resulting spectra (Fig. 7.1) is dominated by ultra-low frequencies. Obviously the annual cycle is the most significant feature, but after that it is the longer periodicities that stand out the maximum period was for a period of 12 years. Various tests were conducted to check that the presence of the dominant long-term periodicities was not a product of the model parameters. While the peak periodicity moves around between 10 and 16 years, the essential result remains the same – large-scale features of the atmosphere exhibit significant periodicities in the range 10 to 40 years.

The significance of this work in the search for weather cycles cannot be understated. First, it demonstrates the central importance of non-linear atmospheric effects. The reason previous simulations had not identified the scale of ultra-low frequency fluctuations was that they did not incorporate this non-linearity. As has been noted in Section 7.1, these consequences of non-linearity will return again and again to complicate matters. Second, the fact that the atmosphere alone can exhibit such long-term fluctuations shows that it is not essential to invoke the slowly varying components of the climate system to explain long-term periodicities. Although these other

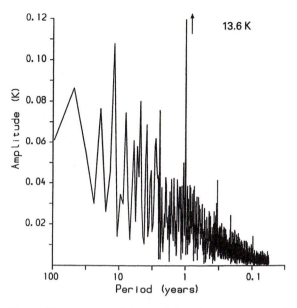

Fig. 7.1. The spectrum of time variation of a temperature-related coefficient in the output of a simple non-linear model of the Earth's climate produced by the Department of Meteorology at the University of Reading, UK. The spectral analysis was carried out on the last 96 years of the model's integration and shows that there is substantial variance for periods longer than 10 years. (From James & James, 1989. With permission of Macmillan Magazines Ltd.)

factors may well prove to be more important, they need to be combined with the natural variability of the atmosphere alone, as we will see in the next section. Third, it places an even greater burden of proof on showing that the observed fluctuations in the weather are linked directly with external influences like solar variability. This is because if a 10- to 12-year cycle were simply the product of natural atmospheric variability, it would undermine those statistical arguments that assume that the observed behaviour had some external cause. The danger is that these attach too much significance to the coincidence between, say, the frequency of the solar variability and the atmospheric variability. In the absence of a good physical argument as to why the one causes the other, the tests of significance must be more searching.

7.3 Climatic feedback mechanisms

The preceding section has put the cat among the pigeons. If all the fluctuations observed in weather statistics are nothing more than a product of the non-linear behaviour of the atmosphere, the search for other physical mechanisms is pointless. But the evidence of the weather records and the description of both the global climate and extraterrestrial influences indicate that this conclusion may be premature. So the idea that the atmosphere can apparently set up long-term fluctuations of its own accord represents a starting point for examining the climate. What we must now do is to examine how the combination of all the climatic components may contrive to produce more pronounced and regular fluctuations. The way forward has already been touched upon in the description both of the atmosphere–ocean model developed for ENSO events and also of the various other ocean–atmosphere oscillations around the world (see Sections 5.5 and 5.7). The ENSO model demonstrated that linear connections between the various components of the wind field over the tropical Pacific combined with the surface properties of the surface layer of the ocean can produce a plausible model of the quasi-periodic behaviour of the ENSO. Changes in one part of the tropical Pacific can take months or years to propagate to other regions, thus making it possible to set up oscillatory fluctuations on these timescales. This type of modelling works most effectively in the relative symmetry of the Pacific basin where the ocean–atmosphere system is amenable to a relatively simple treatment; the observed behaviour cries out for this type of modelling.

In principle, the same type of explanation could apply to other temporary or persistent quasi-cycles. Central to understanding why this should be the case is the nature of the feedback processes that have been mentioned frequently in Chapter 5. These processes arise because, when one climatic variable changes, it alters other variables in a way that influences the initial variable, which triggered the change. If this circular response leads to a re-inforcement of the impact of the initial stimulus then the whole system may move dramatically in a given direction. This runaway response is known as 'positive feedback' and is best illustrated by the high-pitched whistle that a microphone system can produce as it picks up some of its own signal. In the case of the climate, an example might be the effect of a warming leading to the reduction of snow cover in winter. This, in turn, could lead to more sunlight being absorbed at the surface and yet more warming, and so on.

The reverse situation is when the circular response tends to damp down the impact of the initial stimulus and produces a steady state. This is known as 'negative feedback'. An example of this type of climatic response is where a warming leads to more water vapour in the atmosphere, which produces more clouds. These reflect more sunlight into space thereby reducing the amount of heating at the surface and so tend to cancel out the initial warming. It should be added that at the same time more water vapour (a powerful greenhouse gas) will also lead to greater absorption of terrestrial radiation in the lower atmosphere and hence warming: a positive feedback. So, in the case of water vapour we have to consider a combination of negative and positive feedback.

When it comes to modelling feedback processes, much depends on the complexity of the system involved, as these types of chain-reaction can lead to a variety of responses to any perturbations, including sudden switches between different states, and reversals where a positive stimulus can lead to a negative reaction of the system as a whole. Despite this daunting prospect, there is no alternative to using computer models to explore how the atmosphere and the oceans, plus any other external influences we wish to consider, could interact to produce oscillations. Furthermore, the progress made on ENSO gives us some confidence that the same techniques can be applied to the various ocean–atmosphere oscillations described in Chapters 3 and 5.

This confidence faces possibly its sternest test with the North Atlantic Oscillation (NAO) (see Sections 3.7 and 5.7). The combination of weather patterns, ocean currents and the seasonal variation of sea ice in the North Atlantic requires researchers to make much more difficult decisions about simplifying assumptions concerning the dominant factors in the models. Various attempts to simulate the behaviour of the SSTs, the extent of sea ice and atmospheric circulation have not resolved the fundamental issue as to whether the NAO is driven by atmospheric variability or by longer-term changes in the ocean.[4] If anything, it appears that much of the longer-term variability of the atmospheric circulation arises from internal atmospheric processes, although the tripole pattern may exert a significant influence in wintertime. The fact that in winter the phase of the NAO depends on the strength of the stratospheric vortex over the Arctic – when the vortex is strong the NAO tends to be in a positive phase and vice versa – is another interesting feature as this is linked to both the QBO and ENSO events. None the less, there is no doubt that fluctuations in SSTs and the strength of the NAO are related, but it is not clear from model studies whether the atmosphere is driving the ocean or the ocean driving the atmosphere.

A measure of the complexity of these processes is an investigation of the NAO by Carsten Eden and Jürgen Willebrand, of the Institut für Meereskunde, Kiel, Germany, using a model of the Atlantic Ocean, which was driven with decadal-timescale historic atmospheric weather patterns.[5]

---

[4] See for example Rodwell, Rowell & Folland (1999), Robertson, Mechoso & Kim (2000) and Czaja & Frankignoul (2002).

[5] Eden & Willebrand (2001).

The bulk of the variability of the oceanic circulation was directly linked to the NAO. The model showed both a fast (intraseasonal) response and a delayed (timescale about 6–8 years) oceanic response to the NAO. Effectively there was an immediate positive feedback and a delayed negative feedback to the NAO – yet another example of the complicated responses that can occur in ocean–atmosphere models.

Another candidate for such model investigations is the Great Ocean Conveyor (GOC) (see Section 5.9). Andrew Weaver of McGill University, Montreal, E. S. Sarachik of the University of the University of Washington, Seattle, and Jochem Marotze of the Massachusetts Institute of Technology carried out an early example of modelling work in this area, which lifted the lid on this fascinating subject.[6] They investigated the formation of bottom water in the North Atlantic, where much of the world's deep ocean water originates. As explained in Chapter 3, this process is driven by thermohaline circulation in which warm water flows polewards and colder water returns to low latitudes at depth. The controlling factor is how the temperature and salinity vary in the region where the surface water sinks to form deep water. Salinity is the dominant factor in the rate at which polar water sinks and, unlike temperature, this is not subject to local feedback. This is because the salinity is controlled by the balance between precipitation and evaporation, which is not affected by local salinity but by wider atmospheric and oceanic circulation.

Using a simplified model of the North Atlantic, Weaver and his colleagues showed that the thermohaline circulation can exhibit substantial and often chaotic variations. When the model used a distribution of precipitation and evaporation that resembled the climatology of the North Atlantic the changes in circulation occur on a decadal timescale. Even allowing for the simplicity of the model, the fact that it produces quasi-periodic but chaotic fluctuations is fascinating. Subsequent modelling work on both the North Atlantic and the GOC has expanded and refined this early analysis. Although the range of behaviour has inevitably been extended, the central feature remains the repeated prediction of quasi-cyclic multidecadal oscillatory behaviour and the propensity for the circulation of the GOC in the North Atlantic either to vary or to 'flip' between different states. For example, the GDFL model,[7] as noted in Section 5.7, has produced a series of simulations that exhibit multidecadal fluctuations. These involve not only atmospheric–ocean coupling and changes in the strength of the thermohaline circulation but also parallel changes in the extent of sea ice in the

---

[6] Weaver, Sarachik & Marotze (1992).

[7] Delworth (1996) and Delworth, Manabe & Stouffer (1997).

Arctic. Yet again we are faced with the prospect that certain climatic factors can interact in this apparently cyclic and yet unpredictable manner.

The most important of approximately regular variations is of course the QBO. Its strength in the stratosphere, and the ubiquity of its weaker cousin in surface weather patterns, suggests that there could be some general feedback process ensuring that a pattern in one year will be reversed in the next year. Alternatively, the tropospheric changes could be driven by the more regular stratospheric oscillation, or be a product of the combination of both effects. Taking these options in turn, we need to start by considering the possibility of simple biennial feedback mechanisms.

There have been many attempts to propose processes whereby extreme seasons in one year create the right conditions for the reverse to happen in the next year. This could be along the lines of a cold winter setting up the right patterns for a cool wet summer, which sets the scene for a mild wet winter, which in turn leads to a warm dry summer that is followed by a cold dry winter and so on. Short runs of cold and mild winters or good and bad summers lend some support to these theories (see Fig. 1.1). But the fundamental challenge for this simple approach is that we are seeking a process that has a slightly longer mean period between 2.1 and 2.5 years.

If we are trying to explain a strict periodicity of, say, 27 months, we are in real difficulty. This requires a mechanism that, rather than switching winter to winter or summer to summer, moves on by half a season a year – a much stiffer challenge. But all the evidence is that in both the troposphere and the stratosphere the QBO is a much less regular phenomenon. In particular, at lower levels a better approximation may be biennial fluctuations of varying amplitude that every now and then miss a beat. The identified link between the stratospheric circumpolar vortex and the NAO in winter could be part of such a process, all the more so as the former is related to the QBO. This could explain why the effects are stronger at certain times of the year, especially at high latitudes. It might also provide a link between the sustained quasi-periodic oscillation in the stratosphere and the more fleeting changes lower down. It could be that the phase of the stratospheric QBO might trigger or interfere with biennial feedback mechanisms in the troposphere. This might help solve the conundrum of how a weak signal from the stratosphere could influence the far more energetic tropospheric circulation (see Section 3.9).

This approach to the QBO and how it fits into the overall pattern of weather cycles falls into three obvious stages.[8] First, there must be a model for the behaviour of the winds in the equatorial stratosphere. Then, once

[8] For more on the physical explanations of the QBO see Baldwin *et al.* (2001).

this has been achieved, a link must be established between the QBO and the more shadowy examples of quasi-biennial behaviour at lower levels. To complete the picture we need an explanation of how it is connected to other cycles. At present, only the first criterion can even approximately be met. The second is still the subject of considerable debate. As for the whole question of the combined effect of the QBO and other cycles, as the discussion of solar activity in Section 6.1 has shown, the prospects of an answer seem remote.

The mechanism for the quasi-biennial reversal of the winds in the equatorial stratosphere is a good example of how complex atmospheric processes can be. The accepted models involve a combination of processes in the dissipation of upward-propagating Kelvin and Rossby waves in the stratosphere.[9] These waves originate in the troposphere and lose momentum in the stratosphere by the process of radiative damping. This involves rising air cooling and radiating less than the air that is warmed in the descending part of the wave pattern. Westerly momentum is imparted by decaying Kelvin waves in the shear zone beneath the downward-propagating westerly phase of the QBO. Rossby waves perform a similar function in respect of the easterly phase. The eastward-propagating Kelvin waves (Section 5.8) and the westward-propagating Rossby waves (Section 5.1) combine to produce a regular reversal of the upper atmosphere winds. Refinements of this theory involve turbulent processes and also the effects of changes in the absorption of solar radiation caused by alterations in the concentrations of ozone due to the QBO. But it is a measure of the complexity of these explanations that the QBO is not reproduced in standard three-dimensional models of the global climate (GCMs), although special models with additional levels in the region of the tropopause and lower stratosphere are capable of producing a realistic simulation of QBO-like behaviour.[10] So neither the existence of the QBO nor its climatic consequences feature in the models that are the cornerstone of the current estimates of the contribution of human activities on global warming.

The physical complexity of this explanation underlines the scale of the challenges involved in unravelling the cause of the ubiquitous QBO in the troposphere. But two basic features of the stratospheric model do stand out. First, it is not a localised phenomenon, but reflects worldwide interactions. Second, and more important, it is driven from below, as is required by the thermodynamics of the atmosphere. If, however, this feeds back into the troposphere to produce a more general QBO, life gets more difficult. It could just be that the upper atmosphere QBO is a more visible

---

[9] See Holton & Lindzen (1972) and Lindzen (1987).   [10] Takahashi (1999).

manifestation of some more fundamental property of the global atmosphere, which is masked by the continual complex circulation of the troposphere. An alternative is that temperature changes close to the tropical tropopause could affect the strength of convection in the tropics (the phase of the QBO is a significant feature of successful seasonal forecasts of hurricane activity in the Atlantic basin[11]). This could be amplified through the Hadley Cell and the Southern Oscillation and hence the rest of the global circulation, including the depth of the stratospheric circumpolar vortex in winter and hence feed back into the troposphere by modulating the NAO. The fact that intraseasonal oscillations are also linked to ENSO and influence events at higher latitudes (see Section 5.8) may also be part of these processes.

A simpler explanation may be found in the switch between extreme stages of the easterly and westerly phases of the QBO, amounting to some 5% to 10% of the total angular momentum of the atmosphere in the northern hemisphere. Compensating changes at lower levels to conserve angular momentum could result in there being a propensity for different types of circulation pattern to become established. This might lead to extreme seasons being more likely to occur in one phase of the QBO rather than another. So the stratosphere could trigger any propensity for the troposphere to exhibit biennial feedback mechanisms. Furthermore, given the variability of the stratospheric QBO, it is possible that this constructive interference could last for several cycles. In effect, if the periodicity of the QBO is close to 24 months it could lead to an 'excitation' of the tropospheric biennial oscillation, and the two oscillations could become 'entrained' (see Section 7.1). This would lead to periods of biennial oscillation, interspersed with periods when the behaviour of the stratosphere and the troposphere interfered destructively so that no tropospheric cycle is seen. The examples presented in Figs. 1.1, 1.3 and 5.11 are suggestive of this process.

7.4 Extraterrestrial explanations

The combination of variability of the atmosphere and the various feedback mechanisms operating within the climate system may be capable of providing a plausible explanation for most of the observed quasi-periodic features described in this book. But there is no doubt, as seen in Chapter 6, that extraterrestrial influences are capable of affecting our weather. The fundamental question is: are these weak effects capable of producing a

---

[11] See Gray (1990) and Landsea *et al.* (1994).

measurable impact or are they insignificant compared with climatic auto-variance? The evidence presented so far suggests that it is touch and go. So it is important to examine whether under certain circumstances the weather machine can amplify these weak, but clearly periodic, effects.

The key to any explanation is to find a mechanism that enables the small changes due to extraterrestrial influences to trigger more significant shifts in global circulation patterns. As Section 5.1 explained, the weather machine is driven principally by the energy flows at low levels in the atmo-sphere. For this reason, Chapter 5 focused on how fluctuations in certain climatic parameters could alter the energy balance of the climate. But the most impressive examples of periodicities, which appear to be related to solar and tidal effects, seem to be linked to subtle changes in the circulation patterns of the upper atmosphere. So if it is possible to show that minute changes due to these effects can be translated directly or indirectly into shifts in these patterns, then a physical model may emerge.

In spite of the stiff challenge facing any attempt to produce a physically acceptable mechanism for any significant extraterrestrial influence on the weather, there is no shortage of candidates. Some of the subtle approaches have been touched on in Section 6.3. What these show is that small changes in the energy output of the Sun could be magnified in the atmosphere to produce more substantial changes in the Earth's weather patterns. But in all these cases they involve a complicated chain of electrical, magnetic, radiative or chemical effects. While physically plausible, these proposals have not been verified by observation. Moreover, it is hard to see how some, if not all of them, may be readily checked. The natural variability of the weather makes it difficult to verify these effects. Also, the absence of reliable measurements of such a fundamental climatic parameter as cloudiness, let alone past variations in trace constituents and other minor changes in the atmosphere, means that it will be many years before any hypothesis is adequately tested. This is inevitable given that we need to demonstrate that the proposed physical connection is maintained through several cycles.

A similar set of arguments applies to tidal effects on the weather. Although the evidence presented in Chapter 3 shows that a considerable case has been made for a lunar influence on the weather, there is no well-substantiated mechanism for these effects. The most obvious changes in the tidal forces that could influence the weather are the northward and south-ward pull on the mass of the atmosphere, principally in the subtropical high pressure belt. This should show a poleward and equatorward displacement, chiefly in alternate weeks but varying with the longer-term tidal cycles. So the record should show latitudinal shifts of weather patterns and the jet stream, but, although some early studies showed signs of a monthly cycle, in general the evidence of pressure patterns (Section 3.7) does not provide clear examples of multi-year periodic changes that would lend support to these tidal models. Another proposition is that changes in the direction of the tidal force may explain the monthly cycle in rainfall statistics (Section 3.11).

The effects of tidal forces on the oceans can be calculated. The mean slope of the Atlantic Ocean between 45° N and 70° N varies between 6.5 cm upwards when the Moon is at is maximum declination and 6.5 cm downwards at the intervening minimum declination. This could not only affect the strength of the ocean currents but also alter the interchange of water over the submarine shelves and ridges at the entrances to the Arctic Ocean and the Baltic Sea. The warmer saline Atlantic water could lead to changes in ice cover that could have an impact on the weather. Some evidence has been obtained of periodic tidal effects into and out of the Baltic. But the more important changes of the flow of water into the Arctic basin and the outflow of cold deep water have not been monitored adequately to test this hypothesis. Moreover, the tidal effects are likely to be small compared with the fluctuations in the wind fields and, in particular, the variations of the NAO.

The effects on specific ocean currents are likely to be even more complicated. There is evidence that the flow of the Gulf Stream through the Florida Straits responds to the monthly lunar cycle. But longer-term fluctuations have not been measured with sufficient accuracy to identify unambiguously tidal effects. Recent work has, however, estimated the variability of sea level on the offshore side of the Gulf Stream using a wind-forced numerical model. These calculations were combined with data from coastal tide gauges to estimate the variability of the Gulf Stream since World War II. The spectrum of sea level changes at the coast appears to peak at periods of ~150–250 months, which tallies roughly with the 18.6-year tidal cycle,[12] but the broad nature of the spectral feature can hardly be used to claim that that it is the result of lunar effects at work.

Another possible mechanism relates to the dissipation of tidal energy in the deep oceans. By Kepler's laws the recession of the Moon (some 4 cm per year) implies a loss of energy in the Earth–Moon system of some 3 terrawatts, mostly in the ocean. Recent satellite measurements[13] appear to support the hypothesis that the dissipation of tidal energy in the deep sea, and the resulting mixing, are the controlling features of the overall ocean circulation.[14] If, contrary to standard theory, which assumed that the tides dissipated almost entirely by friction in the shallow seas above

[12] Sturges & Hong (2001). [13] Egbert & Ray (2000). [14] Wunsch (2000).

the continental shelf, a considerable proportion was dissipated in the deep ocean, then this could be an appreciable factor in returning deep water to the surface. In this way, the strength of the tides and their variation over time could be a significant factor in the strength of ocean circulation.

7.5   Modelling the ice ages

As discussed in Sections 4.8 and 6.4, the extent to which the succession of the ice ages can be attributed to a limited number of periodicities appears to pose a different set of questions about modelling the link with the Earth's orbital variations. Rather than examining the possibility that some combination of physical effects might amplify weak extraterrestrial influences sufficiently to have measurable impact on the climate, here we seem to be faced with the challenge of which of a number of mechanisms produces the best fit with the observed climatic variations. But, as will now be seen, this is only partially true. The problem is that while the latitudinal and seasonal variations in incident solar radiation due to the precession of the equinoxes (the 19- and 23-kyr periodicities) and the variations of the tilt of the Earth's axis (the 41-kyr periodicity) are probably sufficient to trigger significant climatic changes, the 100-kyr eccentricity periodicity is the weakest of the orbital effects. This poses considerable difficulties as observed changes clearly show that the 100-kyr ice age cycle is the strongest feature of the climatic record in the last 800 kyr. So it is necessary to consider separately the explanation of the observed 19-, 23-, and 41-kyr cycles and the dominant 100-kyr cycle.

A direct approach to possible models of ice ages is the one adopted by the Imbries (see Section 6.4).[15] Instead of using numerical models to test the astronomical theory, they used the geological record as a yardstick against which to judge the performance of various physical models. These models, which have over the years become more sophisticated, fall into two broad categories. The first group adopted an equilibrium approach to the changes in solar radiation. This involved calculating the climatic conditions that should occur for various combinations of orbital parameters. These models were capable of producing realistic changes in temperature patterns, but were inevitably unable to reflect the inertia of the climate system. So the geological record lagged behind any given model's response. The important ingredient missing in these early models was the characteristic timescales

[15] Imbrie & Imbrie (1980).

of growth and decay of the ice sheets, which appear to be of the same order of magnitude as the timescales of orbital forcing.

The alternative approach is to use a differential model in which the rate of change of the climate is a function of both the orbital forcing and the current state of the climate. Not only is this a more realistic representation of climatic behaviour but it also contains a non-linear relationship between the input and the output that has important physical consequences. But it remains controversial, as there is no agreement as to which climatic factors should be given particular emphasis and what values should be attached to them. For this reason the empirical approach used in the model developed by the Imbries will be considered first, as it provides valuable insights into the processes involved.

The Imbrie model adopted a simple approach. It considered only the link between the orbital forcing function and the land ice volume. This model explored the sensitivity of the changes in ice volume to the time constants for the growth and decay of the ice sheets and the lag between the changes in solar radiation falling in summer at high latitudes of the northern hemisphere. The reason for this basic approach was twofold. First, the change in ice volume as recorded in the oxygen isotope records from deep-sea cores is the most accurately defined climatic parameter over the last million years or so. Second, the cryosphere is the part of the climatic system whose characteristic timescales of response closely match the dominant 100-kyr period of the orbital forcing. Indeed, the common feature to emerge from all the modelling work in recent years is that only when the northern ice sheets exceeded a critical size did the 100-kyr cycle take over in the climate equation. This is seen as the key to explaining why before around 800 kya this cycle did not feature strongly in the climate records. The earlier interval of the Cenozoic ice ages, from, 2.4 to 0.8 million years ago (Mya), was almost completely dominated by the 41-kyr tilt cycle. Prior to this, the 19- and 23-kyr cycles were more important. The advent of the 41-kyr cycle around 2.4 Mya is linked with the start of major northern hemisphere glaciation. It has been argued that two factors were responsible for this shift. First, the movement of the continents led to the closing of the Isthmus of Panama around 3 Mya, which altered ocean currents, transporting more moisture to high latitudes in the northern hemisphere and enabling ice sheets to start developing. Second, the rapid tectonic uplift of the Himalayas and parts of western North America during the past few million years has increased the sensitivity of the global climate to orbital forcing. The link between the meandering of the jet stream, these mountainous areas (see Section 5.1) and the regions where the northern ice sheets developed may be a key to the current pattern of ice ages.

If the size of the northern ice sheets is critical, the important factor is the sensitivity of the ice volume to the time constants for the growth and decay of the ice sheets and the lag between the changes in the solar radiation falling in summer at high latitudes of the northern hemisphere. Once the model had achieved a reasonable representation of this long-term behaviour, the more rapid response of the rest of the global climate to the other orbital parameters, which have a linear impact, could be added in to build up a better picture of the progression of the ice ages.

In spite of developments in recent years, the Imbrie model remains a good start to considering the physical processes that may be at work in the 100-kyr periodicity. In this model the parameters were tuned to achieve the best fit between the calculated ice-volume changes and the oxygen-isotope record. The most important features of the model include, first, the orbital forcing being fixed by the changes in the tilt of the Earth's axis and the precession of the perihelion of the orbit. The changes in eccentricity (i.e. the 100-kyr and 413-kyr periodicities) do not exert a significant influence on the seasonal and latitudinal variations of the radiation input. So the orbital forcing used in the model contained only the 19-, 23- and 41-kyr cycles, in spite of the fact that the ice-volume curve over the last 800 kyr is dominated by the 100-kyr cycle. The second essential feature is that the time constants of growth and decay of the ice sheets are markedly different. This reflects the evidence that during the last 800 kyr or so the ice sheets built up slowly, but collapsed dramatically at the end of each ice age.

The best results were obtained with a set of parameters that included time constants of growth and decay of the ice sheets of 42.5 and 16 kyr respectively, and a lag of 2 kyr between the orbital forcing and the response of the climatic state. This produced a good fit over the last 150 kyr (Fig. 7.2), although prior to this the match was less good, especially in the controversial period 350 to 450 kya (see Section 7.6). More importantly, the calculated changes in ice volume included 100- and 413-kyr periodicities, although the relative strengths were wrong, with the former being too weak and the latter too strong. But the fact that these essential periodicities were present at all highlights an important aspect of non-linear models. This is that the output spectrum has major features that are absent in the input forcing function. This is an example of the phenomena described in Section 7.1, where a simple non-linear system can generate sum and difference frequencies. The important feature here is that the choice of the time constants of ice growth and decay plus the non-linearity of the model combine to produce the required longer periodicities. The difference between the 19- and 23-kyr will produce a 110-kyr periodicity and that between the 23-kyr and 41-kyr will produce a 52-kyr periodicity, and the difference between these two is a

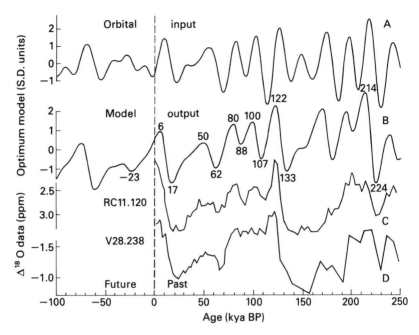

Fig. 7.2. The combined orbital effects shown in Fig. 6.12 can be used as the input (A) to a model whose output (B) shows a marked similarity to the oxygen isotope variations observed in deep-sea cores from the southern Indian Ocean (C) and the Pacific Ocean (D). (From Imbrie & Imbrie, 1980.)

100-kyr cycle. Tuning the model and highlighting a given frequency is both rewarding and also underlines the limitations of the approach adopted. The dependence on empirically derived time constants, which have only the broadest links with the physical behaviour of the ice sheets, is a major limitation.

Clearly, this modelling approach sweeps a lot under the carpet. As an indication of just how complicated the real world may be, analysis of the ice cores obtained at Vostok shows that the CO_2 content in the air trapped in the ice exhibits a remarkable parallelism with the inferred changes in temperature. As Fig. 7.3 shows, in the depth of the last ice age the CO_2 content of the atmosphere was appreciably lower than during the preceding interglacial. Since these changes in CO_2 content would have appreciable climatic consequences, it appears that the changes resulting from the orbital variations were amplified by changes in the atmosphere. This result in no way undermines the basic theory that orbital forcing produced the ice ages, but merely shows that, not unexpectedly, the overall response of the global climate is much more complicated than the simple models considered here might lead us to believe.

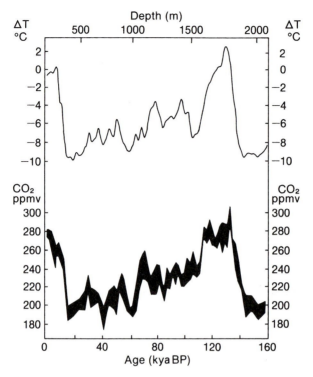

Fig. 7.3. A comparison between (bottom) the CO_2 concentrations in the ice core obtained at Vostok, Antarctica, and (top) estimated changes in temperature over the last 160 000 years. The estimated temperature changes are based on measured deuterium concentrations. As can be seen, the temperature and CO_2 changes march hand-in-hand. (From Barnola *et al.*, 1987. With permission of Macmillan Magazines Ltd.)

The process driving the 100-kyr cycle remains the key challenge in improving ice-age models. An overall assessment produced in 1993 of the many models developed to address this issue confirmed that the non-linear response of the global climate to the northern ice sheets exceeding a critical value was the explanation of the 100-kyr cycle.[16] All that was required was that the build-up of the ice sheets introduced a lag of some 15 kyr into the system for the climate to cease to be dominated by the 23- and 41-kyr cycles and switch into the 100-kyr mode. Components of the atmosphere–ocean–ice system could be responsible for generating this response. The eccentricity cycle need play no part in these changes: they could be nothing more than a natural response of the climate to the variations in the size of the ice sheets.

[16] Imbrie *et al.* (1992) and Imbrie *et al.* (1993a).

An interesting refinement to this type of thinking has recently been proposed by Didier Paillard, at CEA/DSM in France.[17] He produced models which include the possibility of the climate system switching between three distinct climate regimes (e.g. interglacial, mild glacial and full glacial). These different states in the climate system could well be the product of different modes of global ocean thermohaline circulation (see Section 3.7). The switch between these regimes is controlled by a combination of changes in insolation and/or ice sheet volume. By defining the conditions for the transition between the three regimes the model is capable of reproducing with remarkable accuracy the changes in ice sheet volume over the last million years, including the glacial/interglacial cycle between 350 and 450 kya (see above and the next section). So, yet again, the capacity of the global climate to move between markedly different, but relatively stable, states appears to be an essential feature of explaining bigger climate changes. The really interesting challenge that emerges from this work is the need to determine the critical thresholds for the changes between different states and what this means for future climate change. As we will see, however, in Section 7.7 the answer to this question may lie in the stochastic nature of the climate, as transitions between different states may be unpredictable.

7.6 Devils Hole: a contrary view

Although the broad consensus supports the Milankovitch theory of the ice ages, a number of doubts have been expressed as to its ability to explain a number of features of the geological record. In particular, results obtained from Devils Hole – a tectonic cave in Nevada, USA (see Section 4.9) – the walls of which are coated with dense vein calcite (*speleothem*). The stable isotopic content of the calcite provides a record of variations in temperature and other climatic parameters between 60 and 550 kya.[18] Using a combination of uranium-series dating, which provides an absolute timescale, and $\delta^{18}O$ measurements, it produced an estimate of temperatures in the region that corresponds in timing and magnitude to SSTs recorded in ocean sediments off the California and Oregon coasts. The record is also highly correlated with major variations in temperature in the Vostok ice core (Fig. 7.4a), but the correlation with the SPECMAP data is less impressive (Fig. 7.4b).

When the Devils Hole $\delta^{18}O$ time series was published in 1992 it raised awkward questions for the Milankovitch theory concerning the timing of

---

[17] Paillard (1998).   [18] Winograd *et al.* (1988) and Winograd *et al.* (1992).

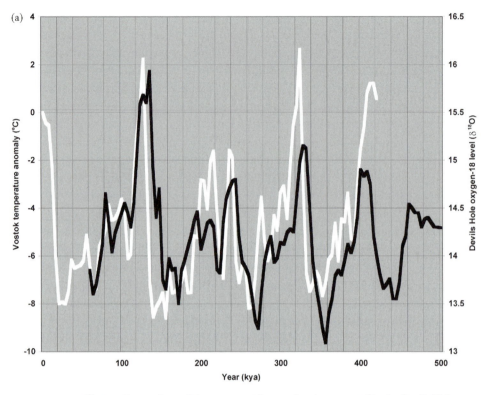

Fig. 7.4. Comparison of the oxygen-18 isotope levels measured in the Devils Hole speleothem (thick black line in (a) and (b)) with (a) the temperature record obtained from the Vostok ice core (thick white line), and (b) the oxygen-isotope record for the SPECMAP compilation of ocean sediment cores (thick white line).

the glacial/interglacial transitions,[19] the duration of the interglacial climates, and the occurrence of a well-developed glacial/interglacial cycle at a time (350 to 450 kya) when insolation theory indicates that none should have occurred. The timing issue now appears to have been resolved. The discrepancy in the duration of the last four interglacials may reflect, in part, differences in the regional nature of the Devils Hole record and the Vostok ice core as compared with the global nature of the ocean sediment cores.

An explanation of these differences may be found in results obtained along the California margin that reveal large (4 to 8 °C) glacial–interglacial changes in SST over the past 550 000 years.[20] This work has analysed both the levels of alkenones and the pollen content of sediment cores drilled in the Pacific stretching down the California and Baja coast to obtain

[19] Imbrie, Mix & Martinson (1993b).   [20] Herbert *et al.* (2001).

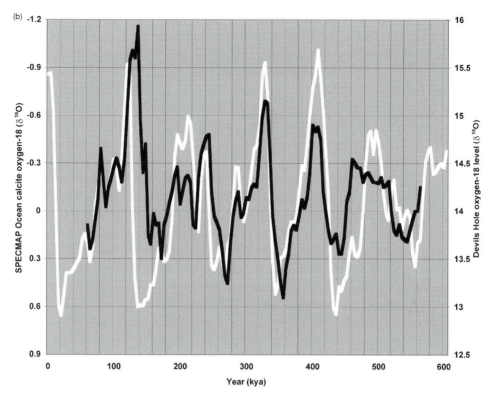

Fig. 7.4. (*cont.*)

information on both SSTs and the nature of onshore vegetation. A particularly valuable marker was the peak abundances in the pollen of Redwood (*Sequoia*), the distinctive component of the temperate rainforest of the north-west coast of California, in interglacial times, when SSTs were equal to or exceeding that of the Holocene. In the region now dominated by the California Current the study concluded that conditions warmed 10 to 15 kyr in advance of deglaciation at each of the past five glacial maxima. Conversely, south of the modern California Current front SSTs did not rise in advance of deglaciation. It concluded, therefore, that glacial warming along the California margin is a regional signal of the weakening of the California Current during times when large ice sheets reorganised wind systems over the North Pacific. Both the timing and magnitude of the SST estimates suggest that the Devils Hole record represents regional but not global palaeotemperatures, and hence does not pose a fundamental challenge to the Milankovitch theory. This potential reconciliation does, however, leave us with the interesting challenge of explaining how parts of the global climate could be so far in advance of the astronomical driver.

The answer may lie in the observed differences between the timing of various climatic events around the world that have been discussed at various places. In addition, there are real difficulties in expecting events that relate to SST changes to match with those that measure effects relating to the build-up of the great ice sheets during each ice age. Nevertheless, it is likely that this argument has not yet run out of steam.[21]

As for the question of the glacial/interglacial cycle around 400 kya, the refinement of the Milankovitch theory by Didier Paillard (see Section 7.5) seems to provide an answer in terms of the non-linearity of the climate system. An alternative intriguing suggestion comes from Richard Muller at the University of California, Berkeley, and Gordon MacDonald[22] at the University of California, San Diego, who proposed that the 100-kyr glacial cycle is caused not by eccentricity but by a previously ignored parameter: the inclination of the earth's orbit to the plane of symmetry of the solar system. Over the last 600 kyr, the best fit between the SPECMAP data and the orbital inclination is obtained with a shift of around 33 kyr. Of particular interest is the presence of a strong variation in inclination near 400 kya.

The real challenge for this proposal is to explain why this particular aspect of the Earth's orbit should produce a significant impact on the climate. The one plausible mechanism links orbital inclination to the extraterrestrial accretion of meteoroids or dust, which might vary as the Earth moved back and forth through the plane of the Solar System. Such material should be detectable in ice and sedimentary rock in the form of changes in the level of iridium, which is present in higher concentrations in extraterrestrial material. If this hypothesis is correct, then a 100-kyr cycle should be seen in ice and sediment records of iridium. Unfortunately, thus far no significant evidence of such changes has been found.

7.7 Stochastic resonance: chaos illumined by flashes of lightning?

The idea of stochastic resonance may seem counterintuitive. It is a phenomenon whereby the response of a non-linear system to a weak periodic input signal is optimised by the presence of a particular level of noise; in effect, increasing the noise level can make it easier to detect a feeble signal. It has attracted a great deal of attention in recent years, notably in the study of non-linear optics, solid state devices, biological studies, including neurophysiology, and also climate studies. Theoretical studies of this type

[21] Winograd (2002).   [22] Muller & MacDonald (1997).

of behaviour suggest that it is not limited to systems with periodic inputs. Thus, in general, noise can serve to enhance the response of a non-linear system to a weak input signal, regardless of whether the signal is periodic or aperiodic.

The implications of stochastic resonances for weather cycles are potentially profound. Virtually everything we have discussed so far in this book can be broadly defined as relating to faint cyclic signals lurking in impenetrable thickets of noise. The possibility that these feeble signals could become more or less evident in the records depending on the level of noise in the climate system is an intriguing one. What is more, if it is linked with the process of entrainment, which results in the combining of, say, the 11-year solar cycle with the quasi-periodic resonant response of an ocean basin to interannual variations in atmospheric circulation patterns, then we may have the ingredients for a 'theory for everything'.

This is all well and good but is it the key to greater understanding, or is it a matter of, in the words of Oscar Wilde when defining the prose style of a fellow author, simply being 'chaos illumined by flashes of lightning'? The answer to this question appears to be whether we can show that the levels of noisiness of the climate vary sufficiently from time to time to result in periods when the level of noisiness is capable of providing the right conditions to enable small periodic perturbations to have a disproportionate influence on the climate and times when it is not. In this respect, perhaps the most impressive application of stochastic resonance in climate studies relates to an area where there is clear evidence of the noisiness of the climate varying appreciably over time: namely during the last ice age (see Fig. 4.10).

A study by Andrey Ganopolski and Stefan Rahmstorf, at the Potsdam Institute for Climate Impact Research, has explored the application of stochastic resonance to the ice age climate.[23] Working with a global climate model of intermediate complexity developed at the Institute, they have demonstrated that stochastic resonance could be an important mechanism for millennial-scale climate variability during glacial times. In particular, they have used this phenomenon to produce an explanation for both the nature and quasi-periodic behaviour of Dansgaard–Oescheger (DO) events (see Section 4.6). The mechanism is based on a stability analysis of the Atlantic Ocean in that there appear to be two circulation modes: one stable, and the other weakly unstable that lasted several centuries before spontaneously ending. In addition, there was a second unstable mode in which North Atlantic deep water (NADW) shut down completely, but this only

---

[23] Ganopolski & Rahmstorf (2002).

occurred when there was a large influx of freshwater (i.e. a Heinrich event –
see Section 4.6) and is not relevant here.

This model suggests that the glacial Atlantic was an 'excitable system', where a suitable perturbation can trigger a temporary state transition to an unstable circulation mode. In contrast, a warm climate, like the
present Holocene, is 'bistable' and appears to be much less susceptible to
disturbance. The model showed that, during glacial conditions, episodic
warm climatic events resembling DO events could be triggered by small-
amplitude sinusoidal variations in the freshwater flux in the high-latitude
North Atlantic. Crucially, this process exhibited a threshold, not in the amplitude of the variation of the freshwater flux (precipitation plus runoff
minus evaporation) but in the background noisiness of the climate. This
stochastic input consisted of adding a realistic level of white noise to the
freshwater flux. The impact of this noise showed an optimum level corre-
sponded to 25% of the net freshwater flux to the northern Atlantic and to
the magnitude of its interannual variability in the present climate, beyond
which the impact declines.

The importance of this work is that it is reasonable to assume that
during the major climatic shifts that were a feature of the last ice age, the
noisiness of the climate could have changed appreciably over time. So, it
is possible that in the early warmer stages of the ice age, when the warm
mode of the North Atlantic circulation was more stable, there were fewer
and longer-lasting DO events. Conversely, at its nadir during the extreme
cold around 20 kya, when the cold mode of circulation was more stable,
there were fewer DO events. In between, when the climate existed in a
metastable state, stochastic shifts between the two states might easily have
occurred. Rahmstorf has since proposed that the timing of the DO events
is triggered by the 1450-year cycle (see Section 4.8).[24] The fact that this cycle
may be of solar origin[25] means that stochastic resonance between the two
modes of North Atlantic circulation could be related to solar variability.

Another interesting potential application of stochastic resonance is
in providing an alternative explanation of the quasi-periodic and chaotic
nature of ENSO behaviour. This work by a group at Tel Aviv University,
Israel, considered the ocean's equatorial wave dynamics as part of a non-
linear dynamical system driven by random climatic fluctuations.[26] Noise
excitation causes the model to jump chaotically between different interacting oscillations generated by the 'delayed oscillator' features of the
model. The injection of noise induced chaotic dynamics, triggering a warm
ENSO (El Niño) event whenever the model's thermocline depth exceeds a

[24] Rahmstorf (2003).   [25] Bond *et al.* (2001).   [26] Stone *et al.* (1998).

threshold level. A 'stochastic resonance' arises, should noise intensity deviate substantially from optimal, with the result that the regular development and coherence of the ENSO cycle is impeded. The model's noise-induced chaos provides an attractive explanation for the sporadic yet deterministic character of El Niño events. The key to changing noise level may be the stochastic behaviour of the weather, including solar irradiance anomalies due to changing cloudiness, precipitation events and westerly wind bursts (i.e. the MJO – see Section 5.8) and other wind fluctuations. These will induce fluctuations in SST and thermocline depth. Of particular interest are the variations in the interannual and interdecadal nature of intraseasonal oscillations (the MJO). This putative explanation does, however, beg the question as to what causes these lengthy variations of the weather in general and the MJO in particular.

Finally, there have been various efforts to explain the overall behaviour of ice ages in terms of stochastic resonance. Indeed, the earliest paper on the subject related to a theory of the ice ages.[27] More recently, a wavelet analysis of various palaeoclimatic proxy records by Han-Shou Liu and Benjamin Chao, at NASA Goddard Space Flight Center, Greenbelt, Maryland, who compared these spectra with those for the astronomically predicted variations of the Earth's orbital eccentricity, obliquity and precession, and their resultant variations of the incoming insolation.[28] This suggested that stochastic resonance could play a part in the 100-kyr cycle in the occurrence of ice ages.

Whether or not stochastic resonance can help solve the remaining doubts about the Milankovitch theory, what is clear, however, is that even with relatively simple models it is possible to produce a relatively good representation of the behaviour of the ice ages over the last few hundred thousand years. By comparison with the other efforts to explain shorter-term periodicities in the climate this is success indeed. So, although a further work needs to be done both in producing more realistic models and in obtaining better measurements of past climatic changes, there is no doubt that this analytical framework sets standards by which other studies of weather cycles must be judged. It also provides a powerful insight into just how complicated the response of the climate is to even well-defined perturbations. This is a useful reminder of the problems meteorologists face in trying to provide a coherent picture of the fluctuations in the weather. With this in mind, we must now consider what general conclusions can be drawn about the nature of periodic and quasi-periodic fluctuations in the weather.

[27] Benzi *et al.* (1982).   [28] Liu & Chao (1998).

Chapter 8

Nothing more than chaos?

It is a tale told by an idiot
Full of sound and fury, signifying nothing.

Shakespeare (*Macbeth*)

Having explored the most obvious of the possible explanations of the observed cyclic and quasi-cyclic variations in the climate, we must now draw some conclusions. But before doing so there is one final fly in the ointment: addressing the ultimate consequence of non-linearity. Wherever possible the aim of the book has been to attempt to provide manageable explanations of the observed fluctuations by using linear models. In Chapter 7 this approach was extended to take in certain aspects of non-linearity to model observed behaviour, notably in the case of the progression of the ice ages (see Section 7.5). In so doing, it has been necessary to note continually that this approach flirted with the unpalatable fact that what we are observing is effectively unpredictable. Frequently, the adjective 'chaotic' has been used in this context and now we need to put this expression on a firmer footing. To do this we need to look at what has become a fashionable areas of scientific thinking since the 1980s – chaos theory.

8.1 Chaos theory

Chaos theory has attracted much public attention for two principal reasons. First, it seeks to bring some understanding to the fascinating boundary between order and disorder in physical systems. Second, it presents the concepts with an intoxicating mixture of imagery. But at a more basic level it provides some particularly important insights into the search for meteorological cycles. This is to be expected as chaos theory has some of its most fundamental roots in meteorology. Edward Lorenz's work at the Massachusetts Institute of Technology in the early 1960s was published in a seminal paper in the *Journal of Atmospheric Sciences* entitled 'Deterministic nonperiodic flow'. The paper[1] provided a variety of fundamental insights into the predictability of non-linear systems, using a simple set of non-linear differential equations to provide a basic representation of convection processes in the atmosphere. All the solutions were found to be unstable and almost all of them were non-periodic. It then went on to consider the implications of this result for forecasting. Here it reached the most profound conclusion. This was that any physical system that behaved non-periodically would be unpredictable. In effect, the system never returns to precisely the same state and so it will never repeat past patterns. Although the weather may follow broadly similar patterns, and these define the climate, it will never return to an identical state but will map out an infinite variety of states that approximate to the climate.

The consequence of this work is that the weather can never be cyclic in the true sense of repeating a given cycle over some period of time. But this is hardly surprising. All the evidence presented in this book has shown that, with the exception of the dominant annual cycle, such exact cycles do not occur. What is more important is whether this conclusion affects our thinking about quasi-periodicities in the weather. Here, Lorenz's work has less to say. Indeed, his simple model inevitably shows quasi-periodic behaviour, but the apparently regular fluctuations vary in length and amplitude so they never return to same spot or repeat precisely the same pattern. Subsequent developments in chaos theory have produced insight into apparently regular fluctuations of the weather.

The two areas of most direct relevance involve, first, the mathematical studies of non-linear equations and, second, laboratory studies of turbulence. In the mathematics of chaos the important phenomenon is 'period doubling'. To understand the full nature of this process it is best to refer to the standard texts (see the bibliography). Here, we will consider only the essential features that emerge from studying the dynamical system given by the expression:

$$x_{t+1} = kx_t(1 - x_t)$$

This simple system is representative of a wide variety of physical processes where the state at time $t+1$ depends in some non-linear way on the state at some earlier time t, and on some constant k. While this crude equation is

[1] Lorenz (1963).

hardly representative of the weather, its basic behaviour reflects the 'memory' in the climate (see Appendix A.7) and so its behaviour can provide some useful insights into the nature of more complex dynamical systems.

The important feature of this expression is the sensitivity of its behaviour to changes in k. Up to a value of $k = 3$ the behaviour is stable, but between 3 and 4 it goes through a remarkable cascade of transitions, which are highly sensitive to the value of k. At 3.2 it develops an oscillation between successive vales of x. At 3.5 the period doubles to recur with every fourth value of x, and by 3.56 the period has doubled again. This is followed by a rapid set of doublings until at $k = 3.58$ the behaviour is chaotic. Even more intriguing is that at values of k between 3.58 and 4 there are isolated regions of periodicities involving multiples of other integers (e.g. $k = 3.835$ marks the start of period 3 doubling and $k = 3.739$ is the starting point of period 5 doubling).

This extraordinary behaviour of such a simple dynamical system with its islands of order in a sea of chaos is highly instructive in considering the behaviour of the weather over time. Given that so much of meteorology is about non-linear feedback processes, it is probable that at times they may mirror some of the features of this simple dynamical system. The model of ENSO events described in Section 5.5 is a good example of such an interactive process. In general, it might be wise to assume that these processes were chaotic, but, as ENSO events demonstrate, they combine to behave in a roughly periodic manner. Another example this type of response can be found in models of predator/prey relationships (see Section 4.10). Work by Robert May, while at Princeton University in the 1970s, showed that a mathematical model of a single species for a simple ecological system – a single species with non-overlapping generations – could produce a wide range of dynamical behaviour, including period doubling.[2] This meant that under certain circumstances this model was capable of producing population fluctuations of the type illustrated in Fig. 4.15. As with Lorenz's work, this result demonstrated that even with relatively simple models it is possible to produce examples of the chaotic behaviour that can flit between order and disorder. The importance of this dynamical response is that it shows how sensitive many physical systems are to the feedback process as reflected in k.

In the real meteorological world the links between various parts of the climate system will vary in strength depending on the various components of the system (e.g. sea surface temperatures or the strength of the midlatitude westerlies). The time of year and hence the phase of the annual

[2] Pool (1989).

cycle could also exercise additional control over any process. If part of the climate system were in a state which was close to what could be defined as a 'window of order' then we might see it drift in and out of periodic behaviour. Moreover, in this process the conditions might favour different points in the period-doubling chain. This type of behaviour might explain not only why periodicities come and go, but also that the observed spectra tend to involve approximately multiples of the basic periodicity (e.g. the wavelet spectra shown in Fig. 3.8). It may also be part of the explanation of the differing behaviour between the seasons. Not only do the conditions at mid and high latitudes vary substantially between winter and summer, but also in spring and autumn there is less likelihood of the situation settling down long enough to establish reproducible patterns. When combined with the possibility that ENSO events are phase-locked to the annual cycle (see Section 5.4) it is hardly surprising that the evidence of periodicities varies with the seasons, with the best examples occurring in winter and summer, with little of note in spring and autumn.

On much longer timescales, the shifts between the 41-kyr and 100-kyr cycle as the dominant cycle in the ice age periodicities is another example of possibly chaotic behaviour. The changes in the northern land masses both from movement of the continents with the closing of seaways (e.g. the Isthmus of Panama) and the mountain building (e.g. the Himalayas and the Rockies) explain these switches. What they show is that shifts in the circulation patterns of the oceans and the atmosphere can lead the global climate to respond in profoundly different ways to relatively small changes in the seasonal distribution of solar insolation. The fact that the 100-kyr cycle appears to be the product of non-linearity, associated with the build-up of the northern hemisphere ice sheets to a critical level, reinforces this proposition.

The other relevant area of the study of chaos is laboratory work on turbulence performed by Harry Swinney and colleagues at the University of Texas, Austin.[3] This involved the study of the frequency distribution of turbulence in a Taylor–Couette system, which consists of two cylinders, one inside the other. The outer cylinder is fixed and the inner one rotates. Sandwiched between them is a liquid containing finely divided aluminium particles. By examining the frequency behaviour of the turbulence as a function of speed of rotation of the inner cylinder using a laser Doppler velocimeter, it is possible to measure the onset of chaos and calculate the power spectrum of the changing turbulence. Examples of these results can be seen in Fig. 8.1. They show that as the speed increases, the behaviour

[3] Zhang & Swinney (1985).

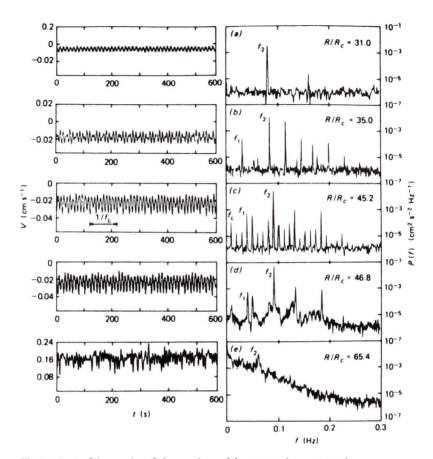

Fig. 8.1. A set of time series of observations of the convection patterns in a Taylor–Couette experiment, and the corresponding sequence of power spectra. As the speed of rotation of the system increases, the spectra become more complicated, moving from a few characteristic periodicities or quasi-periodicities to increasingly chaotic behaviour. (From Thompson & Stewart, 1986.)

becomes increasingly chaotic and the bottom two spectra bear a remarkable resemblance to many obtained from the meteorological series presented in this book. The combination of background noise, with a markedly 'red' distribution in the final example, with a few features whose frequency shifts with the changing speed of rotation looks all too familiar.

Clearly, the parallels between this basic laboratory experiment and the immense complexity of the global climate must be treated with great care. But it does provide a useful insight into the processes at work. Moreover, more complicated laboratory studies of differentially heated rotating systems of both cylinders and hemispheres, often termed 'dishpan experiments', have been able to create realistic analogues of basic global circulation patterns. These have generated Rossby waves in the laboratory, and, as noted in Section 7.2, in a more refined experiment Swinney and colleagues also reproduced the basic form of the Great Red Spot on Jupiter. So the results of the laboratory studies not only confirm the basic form of the onset of chaos in simple systems but they can also model some of the basic features of larger-scale more orderly features of atmospheric motions.

This leaves us in an intriguing position. It suggests that almost everything we have examined could be the product of climatic autovariance. The results of Ian and Paul James (see Section 7.2) may hold the key to almost all the cycles and quasi-cycles paraded here. The theoretical propensity of the atmosphere to exhibit significant periodicities in the range from years to decades can be combined with other more slowly varying features of the climate. In particular, ENSO events demonstrate that the atmosphere and the oceans can interact over several years to produce quasi-cyclic behaviour. The growing array of other forms of oscillations around the world reinforces this conclusion. These atmosphere–ocean interactions, while appearing to represent regional circumstances, are capable of acting in concert or largely cancelling out each other out around the world. This means that they may well be capable of producing much greater fluctuations than is seen in the limited records we have at our disposal.

Modelling work on ENSO and of the thermohaline circulation of the North Atlantic sounds the same warning. It shows how even simple representations of the climate can hover between periodic behaviour and chaos. When various aspects of the climate are braided together in a non-linear way, it is no wonder they are capable of producing many of the observed features of climatic variability. Moreover, the possibility that the climate might either globally or regionally drift in and out of 'windows' of order would be another part of this random model. The recent widespread use of wavelet analysis of time series[4] produces an output that often provides evidence of this behaviour with 'islands' of 'highly significant' periodicities surrounding seas of inconsequence.

This takes us back to the dominant role of the annual cycle. Although this cycle has tended to play a walk-on part throughout this book, the way it interacts with the differing timescales of other fluctuations, both longer and shorter than a year, is crucial to the analysis. In particular, we need to look more closely at how the annual cycle can interfere constructively or destructively with other periodicities. This is of particular importance in the tropics. For instance, the timing between the annual monsoon cycle and the Madden–Julian Oscillation (MJO) influences the strength of the

---

[4] Torrence & Compo (1998).

summer rains of the Indian subcontinent (see Section 3.11). On the longer timescale, we have not unravelled the link between the annual cycle and the QBO in the troposphere: clearly a very real if slightly fuzzy feature of the climate. Similarly, how the development of ENSO events appears to depend on being in phase with certain aspects of the annual cycle requires more work. Furthermore, there is the intriguing impact of this possible phase-locking on the timing of ENSO events, and how this may lead to a Devil's staircase in the period between successive events.[5]

Despite this pervasively chaotic picture, there are a number of features in recent years that suggest that there is still some mileage in the search for cycles. To start with, building on the experience of studying ENSO, knowledge of the general pattern of interannual and interdecadal oscillations is providing new insights into the nature of the global climate. In parallel, in spite of deep wariness on the part of the majority of the climatological community, the case for solar variability exerting a more significant than currently accepted influence on the weather refuses to lie down.[6] The physical arguments will only be resolved when the general issues of the possible links between solar activity, cosmic rays, the Earth's electric field and cloudiness are resolved. In the meantime, much may hinge on the establishment of better statistical links between the periodicity at around 20 years and solar and lunar influences. This scrutiny will have to focus on non-linear interactions between extraterrestrial influences and quasi-periodic autovariance in the climate system, notably entrainment. Alternatively, it could turn out to be a matter of stochastic resonance.[7]

To the extent that there is still life in the theory of the solar origin of the 20-year feature, there is the question of the longer periodicities of around 90 and 200 years. Buoyed up by the growing evidence from a variety of proxy records, the effort described in Section 6.1 to explain the close correlation since 1700 between northern hemisphere temperature trends and solar activity is bound to continue. In particular, this is likely to concentrate on two things. First, we need better measurements of the true nature of longer-term solar variability. This should lead to a more complete theory of true scale of changes in irradiance and the extent to which they can explain observed global temperature variations. Second, and probably even more demanding, we must discover whether there is a link between how various types of cloudiness are changing geographically, and over time, and observed solar activity. Only time will tell whether changes in solar irradiance, magnetic field effects and solar wind, plus associated variations in cosmic ray fluxes

[5] Jiang, Neelin & Ghil (1995).

[6] See for instance Rind (2001).   [7] Rahmstorf & Alley (2002).

and the concentrations of radiatively active trace constituents in the upper atmosphere, together with changes in cloudiness, are all part of this climatic mélange. At the moment, what is clear is that climatic models are in their infancy when it comes to deciding whether these mechanisms could amplify solar variability to produce disproportionate climatic responses.

Whatever the outcome of future research, it is unlikely that it will lead to sufficient understanding to enable long-term predictions to be made of the impact of solar variability on the Earth's climate. Although the sunspot cycle is remarkably regular when compared with climatic fluctuations, as has been made clear in Chapters 3 and 6, it is not possible to predict accurately its future behaviour. When it comes to variations in solar magnetic effects, while the Hale cycle is remarkably regular, longer and shorter variations plus the associated fluctuations in particle emissions are completely beyond our current forecasting abilities. So, even if it transpires that the Sun is playing a more significant part in climate change than majority scientific opinion currently accepts, it is improbable that this would make a substantial difference to our ability to predict future quasi-cycles in the weather.

A similar situation relates to the more specific challenge of solving the riddle of the link between the QBO and sunspot numbers. Ozone changes may be the key to explaining why the phase of the QBO in the stratosphere and the level of solar activity should combine to influence the weather at lower levels. Levels of ozone and other chemical species in the stratosphere vary in phase with the QBO (see Section 6.3). Furthermore, as ozone levels may affect the depth of the polar stratospheric vortex and hence tropospheric circulation patterns at high latitudes in the northern hemisphere in winter, a QBO–solar link could be part of the observed variability. Even if this is part of the story, the next hurdle to clear is the shifting phase of the link (see Section 3.10), which, at times of low solar activity, swings from the polar winter vortex being disturbed and weak when the QBO is easterly and deep and undisturbed when the QBO is westerly, while at times of high solar activity the reverse occurs.

A more radical explanation is that the solar cycle is a red herring.[8] Given that we only have observations since the 1950s, it is not beyond the bounds of possibility that what we are observing is nothing more than a temporary non-linear interaction between the QBO and an interannual natural fluctuation in the troposphere, or even the annual cycle. If so, while the stratospheric oscillation continues unabated, the connection with the troposphere would come and go without warning. This could explain why

---

[8] See section 4.4 of Baldwin *et al.* (2001) for a review of these proposals.

Labitzke and van Loon continue to observe a strong QBO–solar connection in the stratosphere, while the tropospheric winter-forecasting rule broke down in 1988/89 and 1990/91, and then did not reappear with the peak in solar activity between 2000 and 2002. So the best we can say about the QBO–sunspot link is that, while the evidence of effects in the stratosphere remain strong, the jury still out on the case for predictable links with the troposphere.

The other cycle that has been linked to solar variability, and which has been attracting increasing attention in the last decade or so, is the 1450-year periodicity. Its particular importance is in various Holocene proxy records,[9] where it may provide an explanation of both the Medieval Climatic Optimum and the Little Ice Age in the North Atlantic region. It is also implicated in triggering the Dansgaard/Oescheger oscillations during the last ice age.[10] Clearly, this is a cycle that will be worth keeping an eye on in the coming years.

The second major set of cycles that appears to have an indisputable kernel of physical truth is the Milankovitch theory. Clearly, the physical credentials of orbital forcing are good, although the reliance on empirical non-linear behaviour pushes the theory to the edge of chaos and means there is much work to be done. The energetics of the latitudinal solar irradiation variations require some non-linear processes to precipitate the scale of global climate change that has occurred in the past. These mechanisms are, however, easily postulated. The positive and negative feedback mechanisms associated with the expansion and contraction of the northern ice sheets provide plenty of scope for developing models that could amplify the effects of changing solar radiation patterns. So better climate models are likely to be developed in tandem with improved measurements of past climate change.

In addition, the models will need to tackle two more subtle problems. The first, which has already been discussed in some detail in Section 7.5, is to provide an adequate explanation of the dominant 100-kyr cycle. This will involve a better handling of the non-linear interactions of the global climate. The second challenge also relates to these interactions, and is that the palaeoclimatological evidence suggests that the progression of the ice ages was not a smooth process, but came in fits and starts. Long periods of relative stability were followed by rapid shifts in the climate. This behaviour may be the source of the unexplained variance identified in Section 4.8. More important, such sudden changes are symptomatic of chaotic systems. The development of chaos theory may help to unravel more of the mysteries

[9] Bond *et al.* (2001).   [10] Rahmstorf (2003).

of the ice ages. But whatever this work reveals, the underlying fact is certain – the cyclic effects of orbital forcing are the pacemakers for these longer-term climatic changes.

8.2 Future changes

Against this uncertain background there is one remaining question: can we use any of these conclusions about cycles to forecast future weather? The experience of using the QBO–sunspots connection to forecast United States winters in 1988/89 and 1990/91 is good reason to be wary about empirical seasonal forecasts. The fact that it did not receive much attention during the next peak in solar activity is a measure of its fall from favour. In the meantime, ENSO seasonal forecasting had become the flavour of the month and with the success of predicting a very mild winter in 1997/98 its star was in the ascendant. Less impressive results on the back of La Niña event in 1998 and 1999 somewhat undermined this success as both the following winters were even milder than 1997/98. Furthermore, the general difficulties encountered with improving the prediction of ENSO events have reduced confidence in making rapid progress in seasonal forecasting.

The 20-year cycle may have some potential. In particular, the analysis of the drought area indices in the United States (see Section 4.1) made by Murray Mitchell raised some interesting possibilities.[11] But, as he emphasised, the situation was by no means clear, as the supposed complicated interaction between lunar and solar cycles is still not fully understood. At the time, it looked like the lunar cycle had undergone another switch in phase around 1960, in the same way as it may have done around 1800. In which case, when combined with the double sunspot cycle, the American West could have been facing a massive drought around the end of the twentieth century. In practice, rainfall across much of the central USA rose during the last two decades of the twentieth century. Nevertheless, much of Colorado, Nebraska, Utah and Wyoming had record low rainfall in the 12 months ending July 2002, although it is too early to conclude that the combination of the lunar and solar cycles provide the key to long-term prediction of drought in the western United States.

To the extent that the 20-year cycle is probably linked to solar activity, forecasts of future sunspot numbers are also of interest. But, as noted in Section 6.1, using spectral analysis of past variations has not been particularly successful. None the less, forecasts continue to be made. In general,

[11] Mitchell (1990).

they predict a decline in solar activity in coming decades with a marked minimum around 2030. Given the past parallelism between solar activity and global temperature trends, this might be expected to counteract the current warming.[12] What is less clear is by just how much this will offset the warming due to the build-up of greenhouse gases in the atmosphere.

In this context it is worth mentioning the prophetic work of Wallace Broecker at the Lamont–Doherty Geological Observatory, which was published in 1975.[13] He used the principal Camp Century cycles (see Fig. 4.11b) to provide a measure of the long-term variability of the global climate. He then combined these figures with estimates of the likely warming due to the rising level of carbon dioxide in the atmosphere. This produced a forecast that the slight cooling trend of the 1960s and early 1970s would be reversed by a marked warming in the 1980s – a prediction that has so far been borne out by events. But in terms of the wider evidence of cycles of around 80 and 180 years, this success has to be viewed with caution. If, as seems likely, these periodicities are related to solar variability, then we must recognise that solar activity has not followed predicted patterns since the early 1970s (Section 6.1). More important is that, while it is reasonable to assume that temperature trends over Greenland may reflect global changes, there is no reliable evidence of 80- and 180-year periodicities in global temperature records. In 1992, I took a cautious view of this success and perhaps the best way to look at it now is in Broecker's own words, which appeared in an article he wrote in *Natural History* in October 2001:

> My prediction was correct, but was it soundly based? Ten new ice-core records – from Greenland, from Antarctica, and from high-mountain sites elsewhere on the planet – are now available. None show Dansgaard's combined 80-year and 180-year cycles. I thus have been inclined to write off the success of my prediction as just a happy accident.

Lurking behind all these observations about the value of several of the more lengthy quasi-cycles for the purposes of forecasting remains the unanswered question about the significance of solar cycles discussed in Section 8.1. Although the case in favour of solar cycles in the climate has strengthened, the argument is by no means settled. While the evidence of significant correlations between solar cycles and various meteorological parameters has risen, the explanations of the connections are, for the most part, not accepted by the majority of the meteorological community. So, although there is no doubt that the small variations of energy output of the Sun over

[12] Lean & Rind (2001).   [13] Broecker (1975).

the course of the 11-year solar cycle will have minor climatic consequences, and the associated changes in ultraviolet radiation may have proportionately greater impact, other mechanisms are too hotly disputed to permit the inclusion of evidence of longer-term periodicities in predictions of global warming.

The issue of whether solar variability has an appreciable impact on global cloudiness is the prime example of why there is little prospect of an early resolution of the question of whether evidence of cycles can provide useful forecasts. Until we have reliable measurements of cloudiness extending over several decades, we will not be able to decide whether hypotheses about how solar variability affects cloudiness hold water. Leaving aside the basic issue of the need to measure accurately how the amount of different types of cloud in various parts of the world changes over time, there is the challenge of unscrambling parallel and possibly related fluctuations in the climate. For instance, ENSO events that occur every few years and major volcanic eruptions (see Section 6.3) will continue to produce effects that will need to be carefully measured to ensure that changes in cloudiness are properly attributed to the right causes.

As for the ice age cycle, the position is clearer. We are, in all probability, sliding into another ice age that will reach its nadir in 23 000 years (see Fig. 7.2). But even here there are two important uncertainties. First, there is no way of knowing whether this cooling will take place smoothly, or in a series of sudden shifts. If it is the former then it will be imperceptible in our lifetimes. If it is the latter there is no way of knowing when a shift will occur, but it is likely to have a major impact when it takes place. More immediate is the question of how much warming will result from anthropogenic activities and the projected build up of 'greenhouse gases' over the next century. The conclusion reached in the IPCC Third Assessment Report[14] was that 'In the light of new evidence and taking into account the remaining uncertainties, most of the observed warming over the last 50 years is likely to have been due to the increase in greenhouse gas concentrations'. In addition, the report concluded that the increase in temperature rise over the period 1990 to 2100 due to anthropogenic causes would be between 1.4 and 5.8 °C. Changes of this order will far outstrip any cooling, resulting in the much slower change in orbital parameters.

These conclusions need to be viewed in the light of the somewhat pessimistic observations made in Section 8.1 about the potentially chaotic behaviour of the climate. When the uncertainties outlined there are combined with those associated with the natural variability of the climate

[14] IPCC (2001), Summary for Policy Makers, p. 10.

(Section 7.2), the difficulties in deciding just how much of the current global warming is natural or the result of human activities may be greater than suggested by the conclusions reached by the IPCC. The computer models used to estimate the size of the human contribution to recent global warming have major limitations.[15] They cannot reproduce the QBO, ENSO or a climatologically realistic representation of blocking. Furthermore, they are not designed to deal with the more exotic proposals about the physical impact of solar variability. This means we treat with the utmost caution estimates as to what proportion of observed climatic variations is due to autovariance and how much is due to external perturbations.

A better understanding of quasi-periodic weather fluctuations is an essential ingredient to dispelling these doubts. As for any sudden 'flip' in the climate, the IPCC Third Assessment Report[16] notes that in respect of the example of thermohaline circulation in the North Atlantic, most models show weakening of the ocean circulation which leads to a reduction of the heat transport into high latitudes of the northern hemisphere. Even in models where the thermohaline circulation weakens, however, there is still a warming over Europe due to increased greenhouse gases. The current projections using climate models do not exhibit a complete shut-down of the thermohaline circulation by 2100.

Then there is the challenge of modelling the intricate web of connections that link the oscillations that are the atmosphere and the ocean basins, and with it getting a better measure of natural climatic variability. At present it looks as though the current coupled ocean–atmosphere GCMs underestimate the variability on timescales from decades to centuries: the periods that are likely to be most influenced by the interplay between the various interdecadal oscillations around the world.[17] Furthermore, in respect of some proxy records, there is a worrying issue of whether these data are themselves understating the variability of past climates (e.g. the 'segment length curse' in dendroclimatology: see Section 4.1). Recent work on carefully selected tree-ring chronologies from 14 sites in the northern hemisphere extratropics suggests that it is possible to preserve coherent large-scale, multicentennial temperature trends if proper methods of analysis are used.[18] This particular work appears to provide support for the warmth of the Medieval Climate Optimum that occurred between around AD 1000 and 1300, which was followed by the Little Ice Age and the subsequent warming since the late nineteenth century. It may be that these

---

[15] IPCC (2001), see Chapters 7 and 8.
[16] IPCC (2001), Summary for Policy Makers, p. 16.
[17] IPCC (2001), Chapter 8, p. 500.   [18] Esper, Cook & Schweingruber (2002).

changes are, in part, evidence of the 1450-year cycle during the Holocene
(see Sections 4.8, 6.3 and 8.1). So, it is possible that tree rings will, in the
future, be used to provide better estimates of natural climatic variability.
Nevertheless, the prospects of getting a more reliable picture of processes
behind climatic variability and what is driving them will probably depend
on improved GCMs. Only when these can simulate the observed cacophony
of interannual and longer-term variability in the climate system can we start
to have confidence that we are getting a reasonable grip on the problem.
The fact is that these models currently do not even produce ENSO-like
events, so they have a long way to go before they can reproduce not only
ENSO, but the NAO, PDO, AAO, etc.

Even with progress we will have to recognise that we live in a non-
linear world where Devil's staircases, flickering switches and stochastic resonance are part of the scenery. In such a world almost anything
could happen, and in all probability eventually will, although, as noted in
Section 1.2, our current climate does not appear to be strongly chaotic. Our
hope has to be that neither natural changes nor any changes that will be
precipitated by anthropogenic activities will push us too far from the relative stability that has been a feature of the climate during the Holocene:
the period that encompasses all recorded history of humankind and more.

As a set of forecasts the selection above does not add up to much.
It does, however, raise the serious question as to whether all the effort
searching for cycles has been pointless. The answer has to be negative. For
although this work has not produced the reliable forecasts many meteorologists were looking for, it has become an integral part of our understanding
of longer-term climatic fluctuations. This will almost certainly provide the
basis for improved forecasting a few months or even a year or two ahead as
we develop an increasingly global approach to such analyses. At the same
time, the improved understanding of natural climatic variability, including
the true extent of extraterrestrial influences, remains an essential component of making a more accurate estimate of what proportion of the current
global warming is due to natural causes.

These wider goals more than justify the growing efforts to identify the
semblances of order in our climate. But this progress will be scant conso-
lation to those enthusiasts who over the years confidently predicted that
reliable patterns would be found and would enable forecasts to made years
or even decades in advance. In truth, almost everything they have studied so
closely over the last few centuries is little more than the random noise of an
immensely complicated physical system – full of sound and fury, signifying
nothing.

Appendix A

Mathematical background

The purpose of this appendix is to provide a simple guide to the basic mathematics that underlies the search for weather cycles. As such it seeks to give only the most elementary features of what is a complex and wide-ranging subject. In addition, it is designed to reflect the fact that there has been a dramatic upsurge in the availability of computer power and statistical packages to carry out a wide range of highly sophisticated analyses on time series. So, it may help to assess the analysis presented in this book, and also provide a 'health warning' on the pitfalls of the uncritical use of powerful programs without looking closely at what can be achieved with what is frequently noisy data. This approach can, however, only go so far. Moreover, it is highly selective in focusing only on the essential elements of analysing time series. Anyone wishing to consider the analysis of time series in greater depth is advised to refer to the statistical section of the bibliography.

A.1 Measures of variability

Any regular measurement of a meteorological variable (e.g. pressure, rainfall or temperature) can be expressed as a time series. This series can be defined as:

$$X(t) = X_0, X_1, X_2, \ldots, X_N$$

where X_0, X_1, X_2, etc. are successive observations of the given meteorological parameter at equally spaced intervals at times O, Δt, $2\Delta t$, etc. The entire series consists of $N + 1$ observations and covers a period

$P(P = N\Delta t)$. Given we are usually concerned with how $X(t)$ varies from the normal, it is standard practice to define the series in terms of variations about the mean value $\overline{X}$, where

$$\overline{X} = \frac{\sum_{n=0}^{n=N} X_0}{(N+1)} \tag{A.1}$$

So the new series $x(t)$ can be defined as

$$x(t) = (X_0 - \overline{X}), (X_1 - \overline{X}), (X_2 - \overline{X}), \ldots, (X_N - \overline{X})$$

or more simply as

$$x(t) = x_0, x_1, x_2, \ldots, x_N$$

where $x_0, x_1, x_2, \ldots, x_N$ are the deviations of each successive observations about the mean $\overline{X}$, and can be either positive or negative.

The variance of any time series is defined as

$$\sigma^2 = \frac{\sum_{n=0}^{n=N} x_n^2}{(N+1)} \tag{A.2}$$

The value of σ^2 is the standard measure of the variability of any set of observations. This variance can be the product of well-understood periodic variations (e.g. the annual cycle) or of other fluctuations, which may or may not be random. If the known periodic fluctuations are removed from the time series (see Section A.6), it is possible to assess the significance of the remaining features in the series in terms of the square root of the variance (σ), which is defined as the 'standard deviation' of the set of observations comprising the series. This can be done in a variety of ways. First, if the fluctuations are considered to be randomly distributed about the mean then observations can be made about the probability of a single extreme event, or run of extreme events, occurring by chance. There is a 32% chance that any observation will be one standard deviation (σ) from the mean, and approximately a 5% chance it will occur 2σ from the mean (Fig. A.1). These figures are widely used by meteorologists to estimate how often extreme events will occur (often defined as the 'return period') and to analyse whether a run of extremes is the product of chance or evidence of a permanent shift in the mean (i.e. a change in the climate). While of considerable importance in meteorology and climatology, this type of analysis is of limited relevance to the study of weather cycles.

In taking the analysis further, the assumption that residual variance is randomly distributed needs to be considered more closely. It only applies in limited circumstances. Two particular criteria are important. First, the series must be *stationary*. This means that there is no significant long-term trend or other form of significant change in the climate during the period covered by

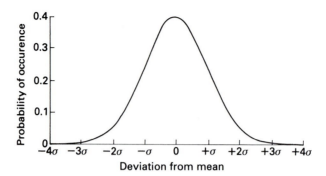

Fig. A.1. The distribution of random fluctuations in a measured variable can be represented as a 'normal' curve, which shows that the probability of any particular value being observed is related to the deviation from the mean. This distribution is usually expressed in terms of the standard deviation (σ), and shows that 68% of observations will be within one standard deviation ($\pm\sigma$) of the mean, 95% will be within two standard deviations ($\pm 2\sigma$) and well over 99% will be within three standard deviations ($\pm 3\sigma$).

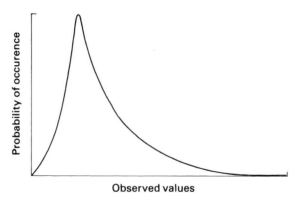

Fig. A.2. With some meteorological variables, the distribution is skewed towards lower values and the scatter cannot be expressed simply in terms of the deviation from the most probable value.

the series. Second, the fluctuations are evenly distributed about the mean. This is a reasonable approximation in terms of temperature and pressure statistics. But in the case of rainfall and wind speed, where the distribution of almost all the figures is markedly skewed toward low values (Fig. A.2), because there are in most records far more observations of, say, zero or low rainfall, than there are of very wet periods. The key to using the best statistical techniques for such series is to establish the statistical nature of the basic distribution of the data. Only when it is clear what form the distribution about the mean takes is it possible to decide what is the appropriate form of statistical analysis to apply to the series.

Returning to extreme events, an aspect that is more relevant to cycles is the analysis of the spacing of such events as defined in terms of being some multiple of the standard deviation from the mean. This approach, which is examined in more detail in Section A.2, is a useful way of checking quickly whether there is something interesting in the distribution of extreme events. It also has the capacity to explore quasi-cyclic behaviour that is sometimes blurred out by complete harmonic analysis.

The third use of the measure of variance is central the study of the products of spectral analysis. To understand how it can be used we first need to consider the basic techniques of spectral analysis (Section A.4) and then consider how random fluctuations in the weather are reflected in time series (Section A.9) and hence affect the spectral analysis. For the moment, all that can be said is that the rules that apply to the significance of extreme features in the time series can be translated in some form to the analysis of the frequency spectrum.

A.2   Sherman's statistic

One useful way of examining the distribution of extreme events is to calculate what is known as Sherman's statistic.[1] If a meteorological time series extending from time d_0 to time d_N contains n events, which exceed a certain threshold (e.g. one standard deviation above normal), and which occur at dates $d_1, d_2, \ldots, d_n$, and so divide the data into $n + 1$ intervals, then the Sherman statistic is given by the expression:

$$\omega = \frac{1}{2D} \sum_{j=1}^{j=n} \left| (d_j - d_{j-1}) - \frac{D}{n+1} \right| \tag{A.3}$$

where D is the total length of the data (i.e. $d_N - d_0$).

The value of the statistic (ω) is a measure of the extent to which each of the actual intervals between the events differs from the average interval, and it varies according to whether the differences in time between successive events are more or less regular than expected from a random series. The probabilities of the value of ω being the product of chance have been calculated and are shown in Fig A.3. This shows that if the value of ω is high, as a result of the extremes coming in bunches, with large gaps in between, then this is unlikely to be the product of chance. Conversely, if the events are regularly spaced, which may be evidence of cyclic behaviour, the value of ω will be too low to be due to chance.

The importance of this statistic is that it is readily calculated. So it provides an easy check on the significance of apparently oddly distributed events and may be used as a way of looking for clues of non-random behaviour in weather

---

[1] See Lamb (1972), p. 235.

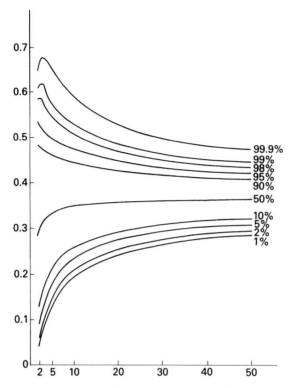

Fig. A.3. The percentiles of the distribution of Sherman's ω-statistic for values of n from 2 to 50. An evaluation can be made of the probability that large values, when events come in bunches with long spaces in between, or low values, when events are spaced regularly, may be the product of chance. (After Craddock, 1968.)

statistics. The Central England temperature series has been fertile ground for such statistical searches. For instance, it has been noted that between 1680 and 1965 the hottest Julys had an unexpectedly high chance of occurring every 13 years or so.[2] Similarly, the coldest Decembers seem to come in bunches at the end of each century.[3] But this type of analysis must be treated with great care. The choice of period covered and the definition of what constitutes an extreme can make a significant difference to the outcome. If we had decided to conduct an analysis in 1974, when Professor Manley published his final series, using the more reliable summer data starting at AD 1700 and considering the hottest high summers (July and August average temperature 17.5 °C or above), we would have concluded that there was good evidence for a 23-year periodicity. The 11 summers identified would have an average spacing of 22.8 years, and a value of the Sherman statistic of 0.172, which has a well below 1% probability of being due to chance. The same analysis carried out for the period 1700 to

[2] Craddock (1968), p. 172.   [3] Burroughs (1982).

2000 would include 15 summers with an average spacing of 18.75 years and a Sherman statistic of 0.22, which is rather less impressive 5% chance. Moreover, the increased frequency of hot summers since 1975 is not limited to ones with values of 17.5 °C or above. If we extend the same analysis to cover those that are 17.25 °C or above we get 18 in the period up to 1974 and 25 in the period up to 2000. The corresponding values of the Sherman statistic are 0.227 and 0.29. The first is a moderately impressive 2% chance, whereas the second is nearer the 10% level. What this shows is that longer-term trends, in this case the marked warming since 1975, can alter the analysis appreciably: in short, if the series has evidence of being non-stationary, as with many other analyses, the value of the Sherman statistic is much reduced.

A.3 Fourier series and Fourier analysis

The underlying mathematical principle of harmonic analysis is that any function, which is given at every point in an interval, can be represented by an infinite series of sine and cosine functions. This series is called a 'Fourier series' and the method of calculating the amplitude of the sine and cosine functions is called 'Fourier analysis'. For meteorological series, observations exist only at discrete points, not continuously. This means that there are a finite number of points over the period of observation and that to make the analysis manageable the observations should be equally spaced. In these circumstances, it can be shown that these points can be analysed in terms of a finite number sines and cosines. For example, if temperatures are given for each of 12 months, five sine and six cosine terms and the mean are sufficient to describe completely the variation over the year. The determination of the finite number of sine and cosine terms is called 'harmonic analysis'. The first 'harmonic' (or fundamental) has a period equal to the total period studied (one year in the example above). The second harmonic has a period equal to half the fundamental period, the third harmonic a period of one third of the fundamental, and so on. In general, if the number of observations is N, the number of harmonics is $N/2$.

The mathematical expression for a Fourier series of some variable over time $X(t)$ is given by:

$$X(t) = \overline{X} + A_1 \sin\left(\frac{2\pi t}{P}\right) + B_1 \cos\left(\frac{2\pi t}{P}\right) + A_2 \sin\left(\frac{2\pi 2t}{P}\right)$$
$$+ B_2 \cos\left(\frac{2\pi 2t}{P}\right) + \cdots + A_n \sin\left(\frac{2\pi nt}{P}\right) + B_n \cos\left(\frac{2\pi nt}{P}\right) + \cdots$$

$$(A.4)$$

where $\overline{X}$ is the average of $X(t)$ over the entire series, n is the number of the harmonic and is an integer between 1 and $N/2$ an P is the fundamental period.

As noted earlier, if $X(t)$ is made up on N observations, there are $N/2 - 1$ sine and $N/2$ cosine terms ($A_{N/2}$ is always zero). So the complete series can be expressed as:

$$X(t) = \overline{X} + \sum_{n=1}^{n=N/2} \left[A_n \sin\left(\frac{2\pi nt}{P}\right) + B_n \cos\left(\frac{2\pi nt}{P}\right) \right] \tag{A.5}$$

which is the sum of all $N/2$ harmonics plus the mean, where each harmonic is defined as

$$h_n = A_n \sin\left(\frac{2\pi nt}{P}\right) + B_n \cos\left(\frac{2\pi nt}{P}\right)$$

Before considering how to calculate the coefficients A_n and B_n there are a number of features about the series that should be defined. First, P is not always equal to N. If observations are made every month for 100 years, $N = 1200$ but $P = 100$ years (i.e. P has units of time whereas N is a pure number). Second, the units of t and P must be the same. So if, in the example just quoted, t is measured in years then $P = 100$, but if t is measured in months then $P = 1200$. On most occasions in this book t is normally measured in years. Finally, the analysis of the harmonics is easier to interpret if the sines and cosines belonging to each harmonic (h_n) are combined. This can be done by adding $A_n \sin(2\pi nt/P)$ and $B_n \cos(2\pi nt/P)$ to form $C_n \cos[2\pi n(t - t_n)/P]$. This means that $X(t)$ is to be redefined as:

$$X(t) = \overline{X} + \sum_{n=1}^{n=N/2} C_n \cos\left[\frac{2\pi n(t - t_n)}{P}\right] \tag{A.6}$$

but the cosine of the difference can be expanded to give:

$$X(t) = \overline{X} + \sum_{n=1}^{n+N/2} \left[C_n \sin\left(\frac{2\pi nt}{P}\right) \sin\left(\frac{2\pi nt_n}{P}\right) \right.$$
$$\left. + C_n \cos\left(\frac{2\pi nt}{P}\right) \cos\left(\frac{2\pi nt_n}{P}\right) \right] \tag{A.7}$$

If this is compared with the earlier expression of $X(t)$ in equation A.5, it can be seen that:

$$A_n = C_n \sin\left(\frac{2\pi nt_n}{P}\right) \tag{A.8}$$

$$B_n = C_n \cos\left(\frac{2\pi nt_n}{P}\right) \tag{A.9}$$

Hence,

$$C_n^2 = A_n^2 + B_n^2 \tag{A.10}$$

$$A_n/B_n = \tan\left(\frac{2\pi nt_n}{P}\right)$$

Therefore,

$$t_n = \frac{P}{2\pi n} \tan^{-1}\left(\frac{A_n}{B_n}\right)$$ (A.11)

Here, C_n is the amplitude of the nth harmonic and t_n is the time at which the nth harmonic first reaches a maximum during the period covered by $X(t)$.

A.4 Calculation of the coefficients of harmonic analyses

In moving on to the calculation of the coefficients of any harmonic analysis of a time series, we must exploit a fundamental property of sines and cosines. This is, given N equally spaced observations at intervals Δt covering the period $P = N\Delta t$, the average of the expression $\sin(2\pi mt/P) \times \sin(2\pi nt/P)$ is zero unless $m = n$. If $m = n$ we must calculate the average value of $\sin^2(2\pi nt/P)$. For values of $n < N/2$ it can shown, using standard trigonometrical functions, that because

$$\sin^2\left(\frac{2\pi nt}{P}\right) = \frac{1}{2} - \frac{1}{2}\cos\left(\frac{4\pi nt}{P}\right)$$ (A.12)

and because the average $\cos(4\pi nt/P)$ is zero, it follows that the average value of $\sin^2(2\pi nt/P)$ is $\frac{1}{2}$. For $n = N/2$ the average of $\sin^2(2\pi nt/P)$ is 1. Also, the average of $\sin(2\pi nt/P) \times \cos(2\pi nt/P)$ over the period P is zero as long as n and m are integers $\leq N/2$.

The consequence of this property of sines and cosines can be seen in the expansion of the time series $X(t)$:

$$X(t) = \overline{X} + \sum_{n=1}^{n=N/2} A_n \sin\left(\frac{2\pi nt}{P}\right) + \sum_{n=1}^{n=N/2} B_n \cos\left(\frac{2\pi nt}{P}\right)$$ (A.13)

If both sides are now multiplied by $\sin(2\pi nt/P)$ and averaged over all N times of the observations, all the terms on the right-hand side of the equation disappear except the one with the coefficient A_n. Hence

$$A_n = \frac{2}{N} \sum_{n=1}^{n=N/2} X(t) \sin\left(\frac{2\pi nt}{P}\right)$$ (A.14)

Similarly, multiplying both sides of the series by $\cos(2\pi nt/P)$ we get

$$B_n = \frac{2}{N} \sum_{n=1}^{n=N/2} X(t) \cos\left(\frac{2\pi nt}{P}\right)$$ (A.15)

But, as has already been shown (Section A.3), it is more normal to combine $A_n \sin(2\pi nt/P)$ and $B_n \cos(2\pi nt/P)$ to give $C_n \cos[2\pi n(t - t_n)/P]$, where C_n is the amplitude of the nth harmonic and t_n is the time at which the nth harmonic first has a maximum within the time covered by the time series $X(t)$.

There is another reason for working with the coefficients C_n. It is standard practice to consider the proportion of the total variance in the time series $X(t)$, which is represented by each harmonic. It follows from the definition given in Equation A.2 that the variance of the nth harmonic is

$$
\sigma_n^2 = \int_0^P A_n^2 \sin^2 \left(\frac{2\pi n t}{P} \right) \mathrm{d}t + \int_0^P B_n^2 \cos^2 \left(\frac{2\pi n t}{P} \right) \mathrm{d}t
$$
$$
\sigma_n^2 = \frac{A_n^2}{2} + \frac{B_n^2}{2}
$$
$$
\sigma_n^2 = \frac{C_n^2}{2} \tag{A.16}
$$

If σ^2 is the total variance in $X(t)$ (see Section A.1), then σ_n^2 can be expressed as a proportion of this figure. Since the harmonics are not correlated, no two harmonics can explain the same part of the variance in $X(t)$. So the variances due to each harmonic can be added, i.e.

$$
\sigma^2 = \sum_{n=1}^{n=N/2} \sigma_n^2
$$

This property means that the values of $(\sigma_n/\sigma)^2$ for each harmonic can be shown as a function of frequency (n/P). This representation is usually known as the power spectrum (see Fig. 2.7) and it displays how much each harmonic contributes to the variance in $X(t)$. The significance of the features in the power spectrum can then be assessed in terms of the spectral distribution that would be expected if the variance were the product of random fluctuations in the weather (see Section A.7).

A.5 Maximum entropy spectral analysis (MESA)

In principle, the computation of the power spectrum of any time series using Fourier transform methods should provide all the frequency information available in the series. This is only true, however, with an infinitely long series. With a finite series, the lack of information about the behaviour of the series outside the period of observation (P) imposes limitations. This is not just a matter of not knowing what occurred before observations started or of what happens after the series ends. It is also the question of the mathematical consequences of working with a truncated set of measurements. This truncation means that the Fourier analysis is a combination of the Fourier transform of the time series plus the Fourier transform of the function that has a value of unity during the period P and zero at all other times. Thus, the computation of any harmonic in the time series is convoluted with the transform of this sampling function. It can be shown that the transform of this sampling function is

$$
W_n = \frac{\sin 2\pi n P}{2\pi n P} \tag{A.17}
$$

The effect of convolving this function with each harmonic of the power spectrum is to produce confusing sidebands. As will be seen in Section A.6, this effect has identical consequences to using an unweighted running mean, albeit on a narrower frequency scale, as the whole time series is effectively 'unweighted'. Moreover, the solution to this problem is similar to the use of a weighted running mean to smooth the time series. This is to give less weight to the beginning and end of the series when computing the Fourier transform. This can be done in a variety of ways, as with running means, but the effect is the same: namely, in removing the problems of truncation of the time series some of the available information is discarded. This is frustrating if the time series is relatively short and there is a need to extract the maximum amount of information from the data.

MESA offers an entirely different approach to this problem.[4] To understand how it works, however, we need to consider the information content of any meteorological time series. This is done by considering the probabilities associated with each observation. If the series contains N observations (x_1 to x_N), we can define the probability of any data point as having a value x_n as being p_n. If all the points were equal, this would tell us nothing special about the weather. The more the observations (and hence the probabilities) vary, the more information is available about the weather. It is possible to define the information content of any information as:

$$I = k \log(1/p_n) \tag{A.18}$$

So the total information content of the times series is

$$I_{\text{total}} = k\{p_1 P \log(1/p_1) + p_2 P \log(1/p_2) + \cdots\}$$

because the chance of observing any particular observation x_n is its probability times the period of the observation (i.e. $p_n P$).

The average information in any unit time interval is termed the 'entropy' (H) where

$$H = I_{\text{total}}/P = -k \sum_{n=0}^{n=N} p_n \log p_n \tag{A.19}$$

and so H is a measure of the uncertainty described by the set of probabilities in the time series. In other words, it is a measure of the ignorance about the precise behaviour implicit in the time series.

The problem addressed by MESA is how to extend the time series effectively to make full use of the available information without adding or taking away information. This leads to the Jayne's Principle of Maximum Entropy, which states:

---

[4] Ulrych & Bishop (1975).

The prior probability assignment that describes the available information, but is maximally non-committal with regard to the unavailable information, is the one of maximum entropy.

This principle can be used to extend the available time series using an indirect method, based on the transform of an autoregressive that neither adds nor subtracts information from the data. The resulting extension of the original data series means that the method is capable of higher resolution than other methods of spectral analysis. Clearly this process can only be extended so far. As shown in Fig. 2.8, beyond a certain limit the increased resolution is only sharpening up the noise and not providing any additional information. There are, however, no precise rules about what is the limit of this process, but practical tests suggest a broad rule of thumb. This is that the series should not be extended by more than $N/3$ for N less than 100, and a decreasing fraction of N should be used as N increases beyond 100. In practice, as Fig. 2.8 shows, this produces only a modest improvement in the analysis of time series where there is only limited evidence of periodicities.

The basic conundrum is that while in theory MESA is designed extract the maximum amount of information from a time series, in reality what is contained in such series does not lend itself to so simple an analysis. The underlying assumption is that the series is effectively made up of signal but no noise, in that there is useful information in all of the harmonics. In these circumstances, MESA can be used to expand the series and squeeze additional information out of the data. By comparison, the basic Fourier transform methods do not attempt to attach too much weight to noise. So the balance of advantage depends on the signal-to-noise ratio. But as is apparent throughout this book, few meteorological series exhibit high signal content.

A greater challenge is that although MESA is supposed not to assume anything about the behaviour outside the range subjected to analysis, the very act of using the randomness within the series to extend the analysis is a major assumption. This approach comes up against the fundamental problem of non-stationary behaviour that is the basic obstacle over which all forms of spectral analysis seek to clamber. Recent alternative approaches, such as wavelet analysis and SSA (see Section 2.8), are examples of this continuing search. In the case of MESA, the examples of the sharpening up of the spectral analysis are often confounded by the subsequent behaviour of the phenomenon under investigation (e.g. the sunspot analysis described in Section 6.1).

This means it is important not to be carried away with the presentational attractions of MESA. Where a complete spectrum is given and the consequences of extension are shown, there is little prospect of being misled. But where only a segment is shown, the emphasis on a single feature may be misleading. So, as a general principle, it is important to have the complete power spectrum so that the contribution of any feature to the overall variance is open to inspection. If at the same time the expected noise spectrum and significance levels (see Section A.7) are shown, the chances of being lured into overweighty conclusions about a single frequency component are greatly reduced, and the value of MESA can be exploited to get a little bit extra out of the data.

A.6 Smoothing and filtering

Although the scope for analysing time series by smoothing and filtering is considered before spectral analysis in Chapter 2, mathematically it makes more sense to take it in reverse order here. The reason for the switch is that while, in a practical sense, the calculation of the running mean of a time series appears easier to do than computing the power spectrum, the mathematics of the process is best understood in terms of its impact on the harmonics making up the series.[5]

The effect of forming any running mean of a time series is understood by showing how a given harmonic h_n (see Section A.3) is modified by the averaging process. This can be calculated in terms of how the nth harmonic (h_n with a period T_n, where $T_n = P/n$) and is modified by a running mean that has a period τ. To do this it is easiest to consider a general $(2K+1)$-point running mean operating on the harmonic h_n. If the sampling interval is Δt, the period of the running mean is $\tau = (2K+1)\Delta t$ and the period of the harmonic becomes $T_n \Delta t$. The effect of the running mean on h_n is to produce a smoothed harmonic H_n where

$$H_n = \left\{ \sum_{k=-k}^{k=+k} f_k h_{n,k} \Big/ \sum_{k=-k}^{k=+k} f_k \right\} \tag{A.20}$$

where f_k is the weight given to the kth sampling point in the range and $h_{n,k}$ are the values of the harmonic h_n over the range of the sampling points. So the value of H_n can be calculated in terms of the values of f_k and $h_{n,k}$. If we limit analysis to symmetrical running means (i.e. $f_{-k} = f_k$) and $f_0 = 1$, then it can be shown that

$$H_n = F_k h_n \tag{A.21}$$

where

$$F_k = \frac{1 + 2 \sum_{k=-K}^{k=+K} f_k \cos\left(\frac{2\pi k \Delta t}{T_n}\right)}{1 + \sum_{k=-K}^{k=+K} f_k} \tag{A.22}$$

---

[5] Burroughs (1978).

So the impact of the running mean is simply to multiply the original harmonic (h_n) by the factor F_k at each point in the time series. F_k is called the filtering function and H_n is the smoothed (or filtered) harmonic.

For the case of the unweighted $(2K + 1)$-point running mean (i.e. $f_k = 1$ for all points), it can be shown that

$$F_k = \frac{\sin\left[(2K+1)\dfrac{\pi \Delta t}{T_n}\right]}{(2K+1)\sin\left(\dfrac{\pi \Delta t}{T_n}\right)} \qquad (A.23)$$

If $\Delta t \ll T_n$ and $K \gg 1$, this can be reduced to

$$F_k = \frac{\sin\left(\dfrac{\pi \tau}{T_n}\right)}{\left(\dfrac{\pi \tau}{T_n}\right)} \qquad (A.24)$$

which is usually called $\mathrm{sinc}(\tau/T_n)$. The form of this function is shown in Fig. A.4. It can be seen that the effect of this function is to filter higher harmonics (i.e. when $\tau > T_n$). But some of these harmonics are not strongly suppressed, and, worse still, where $\mathrm{sinc}(\tau/T_n)$ has negative values the original harmonic is 'smoothed' into a component with opposite sign (i.e. its phase is inverted), which can produce misleading effects. Given that this smoothing package is a standard part of statistical programes on PCs these days, it is as well to be aware of this limitation when using these highly convenient systems. The unweighted running mean does, however, have one useful attribute: when $\tau/T_n = 1$, $\mathrm{sinc}(\tau/T_n)$ is zero. So an unweighted running mean can remove all traces of a cycle whose period is precisely the same as the period of the running mean. This is of particular value when removing the annual cycle from series, as it confirms that simple 12-month average of the data will completely suppress this cycle. So the use of annual averages ensures that, in virtually all meteorological series, no misleading effects occur as a result of the real annual cycle.

A weighted running mean is the way to avoid the unwanted effects described above. The simplest form is a triangular function given by

$$f_k = \left(1 - \frac{k}{K+1}\right) \qquad (A.25)$$

The corresponding filtering function is given by

$$F_k = \frac{\sin^2\left[(2K+1)\dfrac{\pi \Delta t}{T_n}\right]}{(2K+1)^2 \sin^2\left(\dfrac{\pi \Delta t}{T_n}\right)} \qquad (A.26)$$

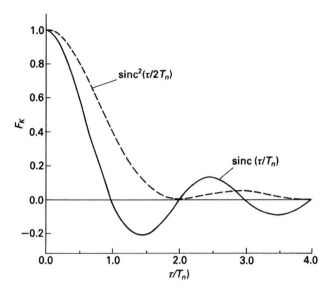

Fig. A.4. The filtering functions for an unweighted running mean (sinc (τ/T_n)) and a triangularly weighted running mean (sinc $(\tau/2T_n)$). The filtering function F_k is the ratio of the amplitude of the harmonics in the running mean to the amplitude of the corresponding components in the original time series. This ratio is shown as a function of the time of interval of the running mean (t) divided by the period of the nth harmonic (T_n) in the time series. (After Burroughs, 1978.)

which for $\Delta t \ll T_n$ and $K \gg 1$ reduces to

$$F_k = \frac{\sin^2\left(\dfrac{\pi\tau}{T_n}\right)}{\left(\dfrac{\pi\tau}{T_n}\right)^2}$$

$$F_k = \mathrm{sinc}^2\left(\frac{\tau}{2T_n}\right) \tag{A.27}$$

As can be seen from Fig. A.4, this is a great improvement over $\mathrm{sinc}(\tau/T_n)$ because

1. the higher harmonics are much more rapidly suppressed (the amplitude of the first subsidiary peak at ($\tau/T_n = 1.5$) is reduced from 0.22 to only 0.045 at ($\tau/T_n = 3.0$)); and

2. the function $\mathrm{sinc}^2(\tau/2T_n)$ is always positive, so even where the filtering is not wholly effective there is no inversion of the higher harmonics.

The disadvantage of the triangular running mean is that its first zero is at (τ/T_n) = 2.0 as compared with (τ/T_n) = 1.0 for the unweighted mean. So to achieve comparable filtering characteristics it is necessary to apply the triangular mean over about twice as many points as the unweighted mean. Indeed, this is a feature of the use of all weighted means. To obtain more efficient suppression of high-frequency components of a time series while leaving the low frequencies largely unaltered requires more points to be included in the computation. This is no longer a problem in terms of exploiting the power of modern computers. It does, however, require careful handling in terms of extracting the maximum amount of information from time series without making unwarranted assumptions about the nature of series beyond the scope of the available observations, especially for relatively short time series. Perhaps the best compromise is achieved using a binomial filter, which can be shown to be a logical extension of the three-point triangular filter. By putting $K = 1$ in equation A.25 we get

$$f_{-1} = f_1 = \tfrac{1}{2} \quad \text{and} \quad f_0 = 1$$

$$F_1 = \cos^2\left[\frac{\pi \Delta t}{T_n}\right] \tag{A.28}$$

We could now carry out this operation K times, at each stage taking the three-point triangular means for the (filtered) component obtained from the previous stage. It follows that after the Kth application, the effect on the original harmonic is such that

$$f_k = \frac{(2K)!}{(K+k)!(K-k)!} \tag{A.29}$$

$$F_k = \left(\cos\frac{\pi \Delta t}{T_n}\right)^{2k} \tag{A.30}$$

The relative merits of this form of smoothing is most easily considered by way of example. Figure A.5 shows the performance of an 11-year binomial running mean (equivalent to $K = 5$ in the iterative application of a 3-year triangular running mean) as compared with a 5-year unweighted running mean and a 7-year triangular running mean (these examples are chosen to smooth many annual meteorological series). As can be seen, for periods longer than 10 years they behave virtually identically. For shorter periods the binomial weighting behaves in a more manageable way as it monotonically declines with shorter periods and removes virtually all variance with a period less than 5 years. The alternative that is now widely used because of the availability of computer power is to use a Gaussian ('normal distribution') weighting. This produces virtually identical results as the binomial and has the fundamental attraction when combining this form of smoothing with Fourier analysis that the Fourier transform of a Gaussian is another Gaussian.

It is possible to construct weighted running means that produce a sharper cut-off of higher frequencies. But, this is at the price of including many more terms in the averaging process for little appreciable benefit in terms of meteorological insights. Such carefully constructed numerical filters are of much greater

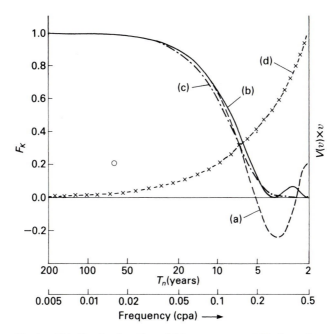

Fig. A.5. This filtering function of: (a) a 5-year unweighted running mean; (b) a 7-year triangularly weighted running mean; and (c) an 11-year binomially weighted running mean. The frequency scale is logarithmic, so to show the impact running means have on 'white noise' at (d), the curve of constant variance per frequency interval (V(v)) is multiplied by the frequency (v) so that equal areas under the curve represent equal variance. (After Burroughs, 1978.)

interest in filtering out both high and low frequencies to explore the behaviour of specific periodicities in time series, in terms of both of their amplitude and how this varies over time.

To understand how the process of narrow-band filtering works it helps to consider what are known as unitary filters.[6] Suppose we divide the frequencies present in the time series into m parts, the range of frequencies is 0 to $\frac{1}{2}\Delta t$, but for simplicity if we assume that Δt is unity then the range is 0 to $\frac{1}{2}$. The unitary filters of the order m can be defined as those filters for which the transmission function has a value unity at only one of the m dividing points or the end points 0 and $\frac{1}{2}$, and zero for all other points. This condition enables the coefficients of the $(m+1)$ filters to be calculated, for the $(m+1)$ frequencies, 0, $1/2m$, $2/2m$, $3/2m$, ..., $(m-1)/2m$, $\frac{1}{2}$. The unitary filter that has a transmission function of unity at frequency $i/2m$ is denoted as $F_{i,m}$ and its coefficients will be $\omega_{0,i}$, $\omega_{1,i}$, $\omega_{2,i}$ to $\omega_{m,i}$. The general formulae for these coefficients are:

---

[6] Craddock (1968), pp. 194–209.

$$\omega_{i,p} = (1/m) \cos \pi i p / m \quad \text{if } i = 1, 2, \ldots, (m-1), \quad \text{and}$$

$$p = 1, 2, \ldots, (m-1)$$

$$\omega_{i,p} = (1/2m) \cos \pi i p / m \quad \text{if } i \text{ or } p = 0 \text{ or } m, \text{ and the other}$$

$$= 1, 2, \ldots, (m-1)$$

$$\omega_{i,p} = (1/4m) \cos \pi i p / m \quad \text{if both } i \quad \text{and} \quad p = 0 \text{ or } m \tag{A.31}$$

To consider how these filters operate it is best to consider an example. Unitary filters of the order 5 have the following coefficients:

$$
\begin{aligned}
F_{0,5} &= (0.100, & 0.200, & \quad 0.200, & \quad 0.200, & \quad 0.200, & \quad 0.100) \\
F_{1,5} &= (0.200, & 0.324, & \quad 0.124, & \quad -0.124, & \quad -0.324, & \quad -0.200) \\
F_{2,5} &= (0.200, & 0.124, & \quad -0.324, & \quad -0.324, & \quad 0.124, & \quad 0.200) \\
F_{3,5} &= (0.200, & -0.124, & \quad -0.324, & \quad 0.324, & \quad 0.124, & \quad -0.200) \\
F_{4,5} &= (0.200, & -0.324, & \quad 0.124, & \quad 0.124, & \quad -0.324, & \quad 0.200) \\
F_{5,5} &= (0.100, & -0.200, & \quad 0.200, & \quad -0.200, & \quad 0.200, & \quad -0.100)
\end{aligned}
\tag{A.32}
$$

and the transmission functions of these filters are shown in Fig. A.6. In practice, these unitary filters provide only a limited suppression of unwanted frequencies, as can be seen in Fig. A.6. If applied to a time series sampled every year, it will centre on frequencies 0.1, 0.2, 0.3, 0.4, and 0.5 cpa (or periodicities of 10, 5, 3.33, 2.5 and 2 years), ignoring filter $F_{0,5}$, which is simply a low-pass filter for frequencies lower than 0.1 cpa. But while filter $F_{1,5}$ will transmit 100% of 0.1 cpa (a 10-year periodicity), it will also let through nearly 50% of 0.167 cpa (a 6-year periodicity), whereas $F_{2,5}$ will transmit all of 0.2 cpa (a 5-year periodicity) and nearly 60% of 0.15 cpa (a 6.7-year periodicity). So the ability of low-order filters of this type to discriminate between intermediate frequencies is limited.

Examination of the coefficients of these simple filters does, however, indicate how more discriminating filters can be produced. The first thing to note is that the sum of all the coefficients is zero for all the filters except $F_{0,5}$, for which the sum is unity. Second, the form of the coefficients of each successive filter is an increasingly rapid oscillation, with $F_{1,5}$ covering less than one cycle, to two full cycles in $F_{4,5}$ (ignoring $F_{5,5}$, which is a high-pass filter centred on the frequency $\frac{1}{2}$, which is rarely of any interest in studying time series). This means that as a general observation the unitary filters are a short segment of a sine or cosine wave of the desired frequency truncated at the start and end of a short interval (i.e. it is the convolution of a sine or cosine wave and the 'box car' function of the unweighted running mean). This means, as can be seen in Fig. A.6, the spectral window of such functions has the $\mathrm{sinc}(\tau/T_n)$ form with its unwanted sidebands. These observations hold the key as to how more discriminating filters can be produced. This will involve combining a few periods of a sine or cosine wave with a symmetrical weighting function of a form such as the binomial discussed earlier in this section. These functions can be used directly

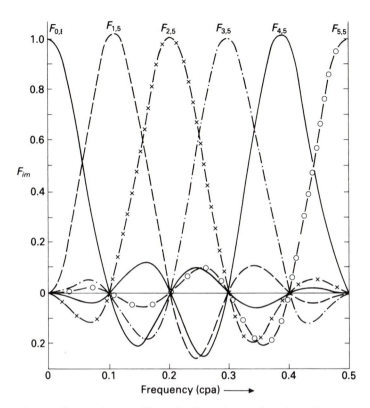

Fig. A.6. The set of unitary filters of order 5, showing how these filters can to a certain extent be used to isolate specific frequencies when smoothing time series.

to analyse the behaviour of certain frequencies within a time series (see Section 4.1) of combined with spectral analysis to explore both the time and frequency dependence of periodic behaviour in times series (see Section A.7).

A.7 Wavelet analysis

Wavelet analysis examines how the power spectrum of a time series varies over the time of the record.[7] Because the technique involves a transform of a one-dimensional time series (or frequency spectrum) to a two-dimensional time–frequency image there is some doubt about the worth of the technique (see Section 2.8). But, as we have seen throughout this book, evidence of cycles and quasi-cycles flits in and out of many time series. As a consequence the Fourier transform of a complete time series smears out these variations, and in the worst case effectively throws the baby out with the bathwater.

[7] Torrence & Compo (1998).

The simplest approach to studying the changing frequency response of some aspect of the climate system is to calculate a form of running Fourier transform of the available time series. This can be done by using a certain window size and sliding it along in time, computing the transform at each time using only the data within the window. This would give us information about the frequency spectrum, but the result will lead to an inconsistent treatment of different frequencies. For a given window width of N sampling intervals in the time series, the width of window would be too small to resolve different low-frequency oscillations, while at high frequencies, although the resolution would be fine, it would be better to have a narrower window to examine the shorter-term time variations of these oscillations.

What wavelet analysis does is to combine both a weighted window and a defined number of oscillations of a given frequency within this window. The convolution of this wave 'packet' for any given frequency provides a measure of the variance of any periodic features within the series corresponding to the frequency range defined by the transform of the window, and how the amplitude variance changes with time. As such it is a version of the band-pass filter described in Section A.6. The choice of the weighting of the window must be chosen to avoid the pitfalls of using a simple 'box-car' form that were identified in Section A.6. The most obvious choice for studying weather cycles is to use a Gaussian envelope, as the Fourier transform of a Gaussian is another Gaussian (cf. Section A.6). The wavelet designed to examine harmonic h_n formed by a plain wave and a Gaussian envelope is given by the expression

$$\psi_n(\eta) = \frac{1}{\pi^{1/4}}\, e^{-\eta^2/2} \cos\left(\frac{2\pi n t}{P}\right) \tag{A.33}$$

where η is a dimensionless constant chosen so that $\psi_n(\eta)$, which is called the 'Morlet' wavelet, covers, say, six periods of harmonic h_n (see Fig. A.7). The Fourier transform of the convolution of this function with a given time series shows how the variance of the frequencies in the vicinity of harmonic h_n change throughout the duration of the series.

The difficulty of using this approach to time series of limited duration is that examination of the longer periodicities is constrained by the length of the wavelet. For example, in the case of the various analyses of the central England temperature (CET) record (see Section 3.1) even the examination of the 23-year periodicity requires a Morlet wavelet containing some 120 terms. This means that only the central part of the record can be examined without making some assumption about the behaviour of the CET record beyond the existing data. For this reason, the presentation of wavelet analyses adopts the convention of showing a cross-hatched area in the bottom corners of the diagrams to indicate a 'cone of influence' where the edge effects become important.

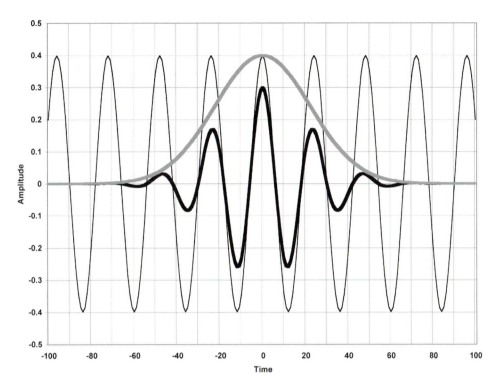

Fig. A.7. A diagram showing how a 'Morlet' function (thick black line) is formed by the combination of a simple harmonic (thin black line) and a Gaussian function (grey line) restricting the influence of the Morlet to about six periods of the harmonic.

A.8 Singular spectrum analysis

Another method of extracting information about temporal variations of power within time series is singular spectrum analysis. This is a variation of the classical mathematical technique of empirical orthogonal function (EOF) eigenvector decomposition that is used in many areas of meteorology to analyse multiple time series at different locations. Instead of using mathematical techniques, which manipulate a matrix of time series from different places to estimate how much of the variance can be attributed to each one of a set of *orthogonal* principal components, it uses a covariance matrix of a single time series with different temporal lags (multiples of a chosen constant time interval). This identifies recurrent time patterns in a single time series. The advantage of this statistical sleight of hand is that it is particularly helpful in isolating anharmonic oscillations with fluctuating amplitudes from noisy data.[8]

[8] Venegas (2001).

This form of analysis has attractions in handling meteorological time series where the signal-to-noise ratio is usually low. By decomposing the data on an orthogonal basis this technique is effectively optimal in a statistical sense. Its limitation is that the length of the time series restricts the choice of the size of the lag and the number of lags that can be used in the analysis (i.e. the width of the moving window that scans the series). This places rather severe constraints on the frequency range over which the temporal structure of the time series can be reconstructed than in the case of wavelet analysis. Typically, with a 100-year annual time series this restriction would limit the analysis to periodicities ranging from around 6 to 30 years. One solution to this challenge appears to be to use a combination of wavelet analysis and SSA.[9]

A.9 Noise

Up to this point in our analysis any reference to the problems of random fluctuations in meteorological time series has not quantified precisely what these mean for the analysis of power spectra. It is now necessary to consider both the spectral consequences of random fluctuations (noise) and how the statistical significance of apparently real features can be assessed. If fluctuations on all time scales were equally probable, then in theory the power spectrum would be a horizontal line (see Fig. A.8(a)). This means that for any unit frequency interval the power density can be expected to be equal. In practice, it is the nature of random processes that the observed variance will show considerable fluctuations with frequency. So, the horizontal line in Fig. A.8(a) is the level for which there is an evens chance of any particular spectral component occurring. In practice, as can be seen in Fig. A.8, there is a considerable scatter about the expected value, reflecting the random nature of the time series analysed and its constituent harmonics. The mean value of the variance in the spectrum is, however, the same as the horizontal line.

On the basis of the figures quoted in Section A.1, 68% of the observations should fall within one standard deviation of the average variance throughout the power spectrum and 95% within two standard deviations of this mean. This suggests that any peak that is more than two standard deviations from the average variance should be regarded as highly 'significant'. But, put another way, even in a purely random time series we could expect to find 5% of the computed power spectrum falling in this range. So, unless there is an a priori reason for any given frequency being present in the power spectrum, the existence of 'significant' peaks cannot be attributed too much physical significance.

These basic statistical strictures must be reinforced by an additional physical caveat when looking for low-frequency cycles in meteorological records.

[9] Yiou, Sornette & Ghil (2000).

(a)

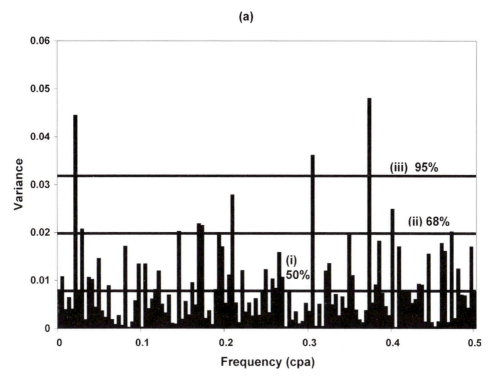

Fig. A.8. Examples of (a) white noise that will not normally be constant as a function of frequency, but will show considerable random fluctuations about the mean variance level (some of which will exceed the 95% 'significance level'), and (b) red noise that shows a marked propensity to have more power at lower frequencies and again show a number of 'significant' features.

This is that the weather appears to exhibit a 'memory' (see Section 2.7). This behaviour reflects the fact that current conditions are influenced by recent events, and, in cases where these influences involve such long-term effects as anomalous SSTs, memory can extend over lengthy periods. This property of the weather can in the first approximation be equated to what is known as a first-order linear Markov process, which can be expressed in the form of

$$x(t) = \beta x(t-1) + \varepsilon(t) \tag{A.34}$$

where $x(t)$ is the observed meteorological parameter at time t, $x(t-1)$ is the same parameter at an earlier time $(t-1)$, β is a constant which represents the serial lag coefficient between successive observations in the $x(t)$ series (where, in general, $0 \geq \beta \geq 1$), and $\varepsilon(t)$ is an independently distributed variable with a mean value of zero and variance $(1 - \beta^2)$ times that of $x(t)$.

This expression implies a power-law (exponential) decay of serial correlation $\beta(t)$ with increasing lag φ, such that for any φ

$$\beta(\varphi) = \beta^{\varphi}; \qquad \varphi = 1, 2, 3, \ldots \tag{A.35}$$

(b)

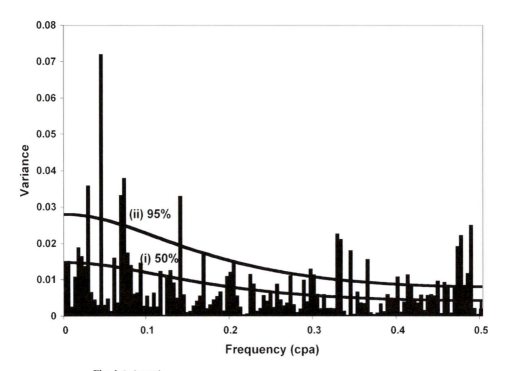

Fig. A.8. (*cont.*)

The corresponding power spectrum of the $x(t)$ series, derived as the cosine transform of this equation, is given as a function of frequency (v) by

$$\Phi v = \frac{1 - \beta^2}{1 + \beta^2 - 2\beta \cos\left(\dfrac{\pi v}{v_{max}}\right)}$$ (A.36)

where v_{max} is the maximum frequency.

If β is small ($\beta \to 0$), this distribution tends towards a white spectrum. If is large ($\beta \to 1$), the spectrum becomes increasingly distorted, with much larger magnitudes at the lower frequencies than the higher frequencies (Fig. A.9), and is usually referred to as a red spectrum. The evidence of meteorological records is that for slowly varying components of the climate system such as monthly SSTs, typical values of β lie in the range 0.5 to 0.9, while for annual figures, including proxy records, the values are from 0.0 to 0.3. So in the analysis of meteorological time series, the significance of features in the computed power spectrum has to be judged on the basis of the underlying variability of the climate exhibiting red noise. This means that the criteria for demonstrating significant low-frequency features are correspondingly more demanding.

In practical terms the handling of 'redness' in power spectra is usually performed by calculating the frequency distribution with the value of β that

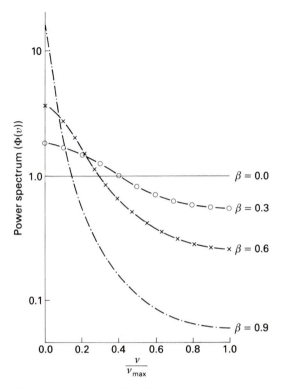

Fig. A.9. The variation of the frequency dependence of random noise in time series as a function of the serial correlation coefficient β as β increases from 0 towards 1.0. With increasing values of β the noise spectrum becomes more and more 'red'.

best fits the general form of the observed power spectrum. This then provides a basis for attributing significance to the most important features in the power spectrum (see for example Fig. 4.4).

A.10 Detrending and prewhitening

There is one other feature of the spectral analysis of times series that can cause difficulties. This is the presence of a trend in the data. This may be the product of real climatic change or simply an artefact of the measurement technique (see Sections 3.2 and 3.3). Whatever the reason, it is preferable to avoid producing misleading effects from changes that are much longer than the period of observation, by removing the trend or using a series in which a trend was never present. Such series are called 'stationary'. As has become clear in many instances throughout this book, meteorological series are rarely ever stationary, so the effects of a trend have to be considered.

For practical purposes the most frequent form of a non-stationary process in meteorological series to a first approximation is a linear trend (other forms of

non-stationary behaviour involving non-linear trends, sudden changes or dis-
continuities can also occur, but do not alter the basic nature of this discussion).
As noted in Section 2.6, the Fourier transform of a linear trend over the period of
a time series produces a power spectrum that is proportional to the inverse of
the square of the frequency. This means that the effect of the linear trend on the
power spectrum is to produce a background that is similar to an extreme case
of red noise. So if the time series is analysed without removing the trend, the
low-frequency elements of the power spectrum will be exaggerated. This will
not produce any spurious features, but it will make it more difficult to assess
the significance of the features that are the product of longer-term variance.
For this reason, it is standard practice to correct for any significant trend before
computing the power spectrum (for example see Section 3.2, Fig. 3.1).

There are two standard approaches to this problem. The first is to remove
the trend ('detrending') in the time series. This is done by computing the underlying linear trend of the observations and then calculating the power spectrum
of the series formed by subtracting the observation at any time t from the value
of the trend at the time. All this does is to replace the mean $(\overline{X})$ in equation
A.5 by the linear trend in the series. The second, known as 'prewhitening', is to
produce a new series by forming the differential of the original series. Strictly
speaking, this is achieved by calculating the midpoint between each successive pair of points in the series and then taking the difference between each
successive midpoint. In practice, an approximation is made in the form of the
difference between successive points in the original series ($x_2 - x_1, x_3 - x_2$, etc.).
Since this is effectively the differential of the original series, it removes both the
trend and low-frequency components while retaining the essential information
about the shorter-term variance.

Although detrending and prewhitening achieve virtually identical results
with meteorological series, in theory they do have somewhat different properties. Detrending is a matter of removing whatever trend is present in the data and
so is determined by the observations themselves. Prewhitening, on the other
hand, is performed by using a pre-ordained formula. This has a specific frequency response and the case of the standard differentiation approach acts as a
high-pass filter of the general form $\sin(\pi n \Delta T / P)$. So the longer-period harmonics (i.e. small n) are suppressed while the higher frequency ones ($n \times \Delta t \to P$)
are left virtually unaltered. In some aspects of time series analysis the difference between these two approaches is significant, notably where there is some
value in removing a dominant low-frequency element from the series. But for
meteorological series these differences rarely assume significant proportions,
and detrending and prewhitening effectively achieve the same purpose.

Glossary

Albedo	the proportion of the radiation falling upon a non-luminous body that it diffusely reflects.
Aliasing	an artifact of the spectral analysis of time series, which occurs when there are significant fluctuations in the measured variable at frequencies greater than that of the sampling frequency, with the consequence that the higher-frequency components are transformed into the computed power spectrum as misleading low-frequency features.
Alkenones	organic chemicals formed from the decay of dead algae that lived in the surface waters that provide an independent measure of sea surface temperature.
Anticyclone	a region where the surface atmospheric pressure is high relative to its surroundings – often called a 'high'.
Autocorrelation	the mathematical process of calculating the correlation coefficient between a time series and the same series with a lag of a number of sampling intervals. The variation of this correlation coefficient as a function of the lag provides information on the existence of periodic fluctuations in the series.
Autovariance	a term used in respect of climatic change to denote the capacity of the global climate to fluctuate of its own accord without the need for extraterrestrial influences.
Blocking	a phenomenon, most often associated with stationary high-pressure systems in the mid latitudes of the northern hemisphere, which produces periods of abnormal weather.

Chromosphere	an irregular layer above the photosphere on the Sun where the temperature rises from around 6000 to 20 000 K.
Coriolis parameter	the Coriolis parameter f is defined by $f = 2\omega \sin ø$ where ω is the angular velocity of rotation of the Earth, and $ø$ is the latitude.
Corona	the tenuous outer atmosphere of the Sun that has an effective temperature of around 1 000 000 K.
Correlogram	a presentation of the degree of autocorrelation in a time series that graphs the autocorrelation coefficient (*see* **Autocorrelation**) as the ordinate and the lag for each coefficient as the abscissa.
Cosmogenic isotope	any species of isotope created by the interaction of cosmic radiation and the atmosphere or the surface of the Earth.
Dendroclimatology	the science of reconstructing past climates from the information stored in tree trunks as the annual radial increments of growth.
Devil's staircase	a mathematical structure that can occur with phase-locking between a dominant driving cycle (e.g. the annual cycle) and a natural oscillation in the climate system (e.g. ENSO), when the frequency of one is not a proper fraction of the other, which leads to chaotic shifts in the frequency of the natural oscillator.
Ecliptic	the great circle in which the plane containing the centres of the Earth and the Sun cuts the celestial sphere.
El Niño Southern Oscillation (ENSO)	a quasi-periodic occurrence when large-scale abnormal pressure and sea-surface-temperature patterns become established across the tropical Pacific every few years.
Feedback mechanism	a process of system dynamics in which a system reacts to amplify or suppress the effect of a force that is acting upon it. For example, in the climate system, warmer temperatures may melt snow and ice cover, revealing the darker land surface underneath. The darker surface absorbs more solar energy, causing further temperature increases, thus melting even more snow and ice cover, and so on. This is positive feedback, in

Feedback mechanism (*cont.*) which warming reinforces itself. In negative feedback, a force ultimately reduces its own effect. For example, when the Earth's surface grows warmer, more water evaporates, forming more clouds. If the clouds which form are extensive and widely distributed, covering large areas of the surface, they will tend to reflect more solar radiation back into space than the dark ground underneath would, cooling the Earth's surface – and reducing the force of warmer temperatures.

Foraminifera an order of Sarcodina, the members of which have numerous fine anastomosing pseudopodia and a shell that is calcareous; the shells of these organisms, when deposited in oceanic sediments, are the source of climatic information.

Fourier transform spectral analysis the mathematical determination of the amplitude of the harmonic components of a time series and the presentation of these in the form of a power spectrum (see **Power spectrum**).

Greenhouse effect an atmospheric process in which the concentration of atmospheric trace gases (greenhouse gases) affects the amount of radiation that escapes directly into space from the lower atmosphere. Short-wave solar radiation can pass through the clear atmosphere relatively unimpeded. But long-wave terrestrial radiation, emitted by the warm surface of the Earth, is partially absorbed and then re-emitted by certain trace gases.

Hadley cell the basic vertical circulation pattern in the tropics, where moist air rises near the equator and spreads out north and south and descends at around 20° to 30° N and S.

Hale cycle the 22-year cycle in solar activity which is a combination of the 11-year cycle in sunspot number and the reversal of the magnetic polarity of adjacent pairs of sunspots between alternate cycles.

Half-life

the time in which half of the atoms of a given quantity of radioactive nuclides undergo at least one disintegration.

Holocene

the relatively warm epoch, which started around 10 000 years ago and runs up to present time. It is marked by several short-lived particularly warm periods, the most significant of which, from 6200–5300 years ago, is called the Holocene optimum.

Interglacial

a warmer period during glacial epochs when the major ice sheets recede to higher latitudes.

Interstadial

a relatively warmer stage within a glacial phase during which the ice advance is temporarily halted.

Intertropical Convergence Zone (ITCZ)

a narrow low-latitude zone in which air masses originating in the northern and southern hemispheres converge and generally produce cloudy, showery weather. Over the Atlantic and Pacific it is the boundary between the north-east and south-east trade winds. The mean position is somewhat north of the equator but over the continents the range of motion is considerable.

Ionosphere

that part of the upper atmosphere in which an appreciable concentration of ions and free electrons normally exists.

Jet stream

strong winds in the upper troposphere whose course is related to major weather systems in the lower atmosphere and which tend to define the movement of these systems.

Kelvin waves

gravity-inertia waves, which occur in the atmosphere and the oceans, where either the effect of the Coriolis Force is negligible (i.e. close to the equator) or where this force is balanced by the pressure gradient. The most important examples are in the equatorial stratosphere and in the thermocline of the equatorial Atlantic and

Kelvin waves (*cont.*)	Pacific close to the equator (in both cases the waves propagate eastwards relative to the Earth).
Last Glacial Maximum	the coldest period at the end of the last ice age between around 24 000 and 18 000 years ago when the ice sheets over the northern hemisphere reached their greatest extent.
Little Ice Age (AD 1550 – AD 1850)	a period marked by more frequent cold episodes in Europe, North America, and Asia, during which mountain glaciers, especially in the Alps, Norway, Iceland, and Alaska expanded substantially.
Maunder minimum	a period during the seventeenth century when the level of solar activity, as reflected by the number of sunspots, was much lower than in subsequent centuries.
Madden Julian Oscillation (**MJO**)	waves of cloudiness, first identified in satellite imagery, that circulate the Earth in the tropics with a period of around 40 to 50 days and exert a considerable influence on tropical weather patterns.
MESA (Maximum Entropy Spectral Analysis)	a method of analysing time series that uses autoregressive methods to extract the maximum amount of information from the available data.
Monsoon	a seasonal reversal of wind, which in the summer season blows onshore, bringing with it heavy rains, and in winter blows offshore — it is of greatest meteorological importance in southern Asia. The word is believed to be derived from the Arabic word '*mausin*', meaning a season.
Non-linearity	the lack of direct proportionality of the input and output of a physical system.
North Atlantic Oscillation (**NAO**)	an index of the circulation in the North Atlantic that is measured in terms of the difference in pressure between the Azores and Iceland. In winter this index tends to switch between a strong westerly flow with pressure low to the north and high in the south and the reverse: the former tends to produce above normal temperatures over

(NAO) (*cont.*)

much of the northern hemisphere, the latter the reverse.

Nutation

oscillation of the Earth's pole about the mean position. It has a period of about 19 000 years and is superimposed on the precessional movement.

Obliquity of the ecliptic

the angle at which the celestial equator intersects the ecliptic. At present, this angle is slowly decreasing by 0.47 arcseconds a year, due to the precession and nutation. It varies between 21° 53′ and 24° 18′.

Photosphere

the visible surface of the Sun on which sunspots and other physical markings appear, it is the limit of the distance into the Sun that we can see.

Plage

bright areas seen around the edges of sunspots that are regions of rising hot gas (so called because they look like sandy beaches around darker islands: plage is French for beach).

Power spectrum

the presentation of the square of the amplitudes of the harmonics of a time series as a function of the frequency of the harmonics.

Precession of the equinoxes

the westwards motion of 50.27 arcseconds per year of the equinoxes, caused mainly by the attraction of the Sun and Moon on the equatorial bulge of the Earth. The equinoxes thus make one complete revolution of the ecliptic in 25 800 years and the Earth's pole turns in a small circle of radius 23° 27′ about the pole of the ecliptic.

Proxy data

any source of information that contains indirect evidence of the past changes in the weather (e.g. tree rings, ice cores and ocean sediments).

Quasi-biennial oscillation (QBO)

the alternation of easterly and westerly winds in the equatorial stratosphere with an interval between successive corresponding maxima of 20 to 36 months. Each new regime starts above 30 km and

(QBO) (*cont.*)	propagates downwards at about one kilometre a month.
Rossby wave	in the atmosphere a wave in the general circulation in one of the principal zones of westerly winds, characterised by large wavelength (*c.* 6000 km), significant amplitude (*c.* 3000 km) and slow movement, which can be both eastward and westward relative to the Earth. In the ocean, similar waves have a wavelength of an order of a few hundred kilometres and nearly always move westward relative to the Earth.
Singular spectrum analysis	a statistical technique that uses the covariance matrix of a time series with different temporal lags to identify recurrent time patterns in the time series.
Speleothem	a deposition of calcium carbonate encrustations by running water, including stalactites and stalagmites, in caves that contain a record of changes in the isotopic ratios in precipitation.
Stadial	a period during glacial epochs when the ice sheets advanced to lower latitudes.
Stochastic resonance	a phenomenon whereby a non-linear system responds to a weak periodic input signal most effectively at a particular level of noise: above and below this noise level the response is less pronounced.
Stratosphere	the portion of the atmosphere, typically between an altitude of 12 to 40 km, where the temperature is approximately constant and there is little or no vertical mixing.
Thermocline	a region in the ocean of rapidly changing temperature between the warm upper layer (the epilimnion) and the colder deeper water (the hypolimnion).
Time series	any series of observations of a physical variable that is sampled at set constant time intervals.

Troposphere	the portion of the atmosphere from the Earth's surface to around 12 km in which temperature falls with increasing altitude.
Variance	the mean of the sum of squared deviations of a set of observations from the corresponding mean.
Varve	distinctly and finely stratified clay of glacial origin, deposited in lakes during the retreat stage of glaciation. Where these stratifications are of seasonal origin they can be used to study climatic change.
Younger Dryas (12.9–11.6 kya)	a sudden, abrupt cold episode, which interrupted the sustained warming trend between the Last Glacial Maximum and the Holocene.

Annotated bibliography

This bibliography is designed to assist the reader to explore in more depth the various aspects of the search for weather cycles and the attempts that have been made to explain observed fluctuations in the weather. It also provides a short set of the more accessible works on statistics which are most relevant to the study of weather cycles. In general these are of historical value, as much of the recent work on solar-weather links, solar variability and the study of long-term ocean–atmosphere variations has yet to be the subject of books.

Climatology and climatic change

Barry, R. G. & Chorley, R. I. (1998). *Atmosphere, Weather and Climate*. London: Routledge.

The seventh edition of a well-established widely read standard work which provides an up-to-date treatment of current meteorological theory and practice with a global perspective.

Bradley, R. S. & Jones, P. D. (eds.) (1995). *Climate since AD 1500*. London: Routledge.

A comprehensive set of papers from leading figures in the climatology world that review the instrumental and proxy records of climate change during the last 600 years. As such it provides a balanced picture of both the evidence of the Little Ice Age and the warming since the late nineteenth century.

Burroughs, W. J. (1997). *Does the Weather Really Matter?* Cambridge: Cambridge University Press.

A book that seeks to provide a balanced and accessible analysis of the current debate on climate change. It combines a historical perspective, economic and political analysis together with climatological explanations of the impact of extreme weather events on aspects of society. As such it provides a background against which to evaluate the potential implications of weather cycles.

Burroughs, W. J. (2001). *Climate Change: A Multidisciplinary Approach.* Cambridge: Cambridge University Press.

This book provides a concise, up-to-date presentation of our current knowledge of climate change and its implications for society. This enables the reader to put the claims about weather cycles into the context of our current understanding of the causes of climate change, both in terms of current changes and those that have occurred throughout the Earth's history.

Diaz, H. F. & Markgraf, V. (eds.) (2000). *El Niño and the Southern Oscillation.* Cambridge: Cambridge University Press.

A series of papers by scientists currently working on various aspects of both the physical nature of ENSO, its impact on the global climate, its history and its social and economic implications for many parts of the world. As such it provides a valuable source of many facets of this global quasi-periodic climatological phenomenon.

Fritts, H. C. (1976). *Tree Rings and Climate.* London: Academic Press.

A standard text by a leading authority on the extraction of climatic information from tree rings. It provides a comprehensive and informative review of many features of dendrochronology. But it concludes that, as of the mid-1970s, there was little evidence of weather cycles in tree-ring data. So it has to be read in the context of more recent developments (see Chapter 4) that provide a stronger case for such periodicities.

Gregory, S. (ed.) (1988). *Recent Climatic Change.* London: Belhaven Press.

A series of papers that review recent evidence of climatic change in various parts of the world, some of which appear to show periodic behaviour.

Grove, J. M. (1988). *The Little Ice Age.* Methuen, London.

An immensely thorough review of the evidence and consequences of the Little Ice Age. It is particularly valuable in that it extends its comprehensive analysis to cover the contraction and expansion of glaciers around the world throughout the Holocene, which provides detailed information about periodic millennial changes in the climate since the last ice age.

Herman, J. R. & Goldberg, R. A. (1978). *Sun, Weather and Climate*. Detroit: Grand River Books.

A thorough review of the various aspects of the evidence of solar variability influencing the climate. It presents a balanced picture of the nature of solar variability, the evidence of long- and short-term climatic change and then considers the physical processes and mechanisms which may link solar variability and climatic change. This is an excellent source of background information on developments up to the mid-1970s.

Imbrie, J. & Imbrie, K. P. (1979). *Ice Ages: Solving the Mystery*. London: Macmillan.

An accessible presentation of the research into the causes of the Ice Ages. It is particularly interesting in providing a personal insight into the work during the 1960s and 1970s that established the modern theory of the ice ages.

Intergovernmental Panel on Climate Change (2001). *Climatic Change: The Scientific Basis*. Cambridge: Cambridge University Press. (For details of the earlier assessment reports, see the references.)

The most comprehensive set of surveys of the evidence of climate change and a detailed analysis of the consequences of anthropogenic activities, including increasing the level of 'greenhouse gases' in the atmosphere, plus forecasts of the likely impact on the global climate over the next century. In spite of the categorical statements made about the contribution of anthropogenic activities to current global warming, the body of this massive and authoritative study contains a bewildering array of caveats about the uncertainties in modelling the global climate, which provide good reason for being cautious about the precise extent of future warming.

Labitzke, K. G. & van Loon, H. (1999). *The Stratosphere: Phenomena, History and Relevance*. New York: Springer-Verlag.

A fascinating and thorough review of the history of scientific studies of the stratosphere, together with detailed analysis of the evidence for quasi-biennial and solar signals in the climatology of the upper atmosphere.

Lamb, H. H. (1972, 1977). *Climate – Present, Past and Future*, Vols 1 and 2. London: Methuen.

The classic work on all aspects of climatic change, which considers the complete range of meteorology, climatology, the evidence of climatic change and possible explanations of observed changes. Of particular interest is that these works devote considerable attention to cyclic aspects of the weather and so provide useful background reading of the position on weather cycles up the early 1970s.

Philander, S. J. (1990). *El Niño, La Niña and the Southern Oscillation*. London: Academic Press.

A thorough and penetrating survey of research into the nature and causes of large-scale climatic changes in the tropical Pacific and their influences on global climate. It is an excellent source of background reading in exploring the nature of quasi-periodic autovariance in the climate.

Trenberth, K. E. (ed.) (1992). *Climate System Modelling*. Cambridge: Cambridge University Press.

A comprehensive set of papers by leading authorities on computer modelling of the global climate that presents many basic insights into the challenges facing scientists wishing to simulate the workings of the climate.

Tyson, P. D. (1986). *Climatic Change and Variability in Southern Africa*. Cape Town: Oxford University Press.

Although concentrating on Southern Africa, this book provides useful analysis of both long- and short-term climatic change in the southern hemisphere, and considers the evidence of cycles in weather records.

Evidence of periodicities

Alley, R. B. (2000). *The Two-Mile Time Machine: Ice Cores, Abrupt Climate Change and Our Future*. Princeton University Press.

A vivid description by one of the foremost researchers in the field of ice-core studies of the challenges of extracting information from the world's major ice sheets and the invaluable information this can provide about past climate change and the potential implications this has for predicting future climatic developments.

Baillie, M. G. L. (1995). *A Slice Through Time: Dendrochronology and Precision Dating*. London: Batsford.

A highly accessible presentation of the basic aspects of dendrochronology that provides an easy introduction to just how much information can be extracted from tree rings.

Pecker, J. C. & Runcorn, S. K. (1990). *The Earth's Climate and Variability of the Sun over Recent Millennia*. Cambridge: Cambridge University Press.

A collection of papers presented at a joint meeting of the Académie des Sciences and the Royal Society held in February 1989, which provides a particularly comprehensive and up-to-date review of the nature and origin of solar variability and a set of interesting observations about how this behaviour may be linked to climatic change.

Rampino, M. R., Sanders, J. E., Newman, W. S. & Konigsson, L. K. (1987). *Climate History, Periodicity and Predictability*. New York: Van Nostrand Reinhold.

A series of papers, prepared in honour of Professor Rhodes W. Fairbridge on the occasion of his seventieth birthday, which provides a comprehensive survey of many aspects of the search for, causes of, and consequences of climatic periodicities. It also contains a huge bibliography of earlier work that is the gateway to much wider reading about the history of searching for weather cycles.

Chaos theory

Gleick, J. (1988). *Chaos: Making a New Science*. London: Heinemann.

An illuminating description of the emergence of chaos theory and the personalities involved in developing the new science. Journalistic in style, this book contains a great deal of interesting material on the behaviour of non-linear systems.

Lorenz, E. N. (1993). *The Essence of Chaos*. London: UCL Press.

An idiosyncratic and insightful description of some of the essential features of chaos theory by the man who made some of the earliest mathematical studies of the unpredictable nature of atmospheric and climatic systems.

Stewart, I. (1989). *Does God Play Dice? The Mathematics of Chaos*. Oxford: Basil Blackwell.

Another popular account of the various components of chaos theory, which concentrates much more on the basic mathematics of non-linear systems. As such, it is of more direct relevance to interpreting some of the quasi-periodic behaviour of the weather discussed in this book.

Statistics

Bloomfield, P. (2000). *Fourier Analysis of Time Series: An Introduction*, 2nd Edn. New York: John Wiley.

An up-to-date version of the well-known textbook that provides an accessible and comprehensive treatment of the mathematics of analysing times series using Fourier transform methods.

Craddock, J. M. (1968). *Statistics in the Computer Age*. London: English University Press.

Although somewhat dated, this book provides an accessible and basic description of the statistical techniques for examining meteorological data. Its emphasis on meteorology is particularly relevant to the issues addressed in this book.

Kendall, M. (1976). *Time Series*. London: Charles Griffin.

A more thorough presentation of the mathematical techniques for analysing the nature and information content of time series.

Panofsky, H. A. & Brier, G. W. (1958). *Some Applications of Statistics to Meteorology*. Pennsylvania State University Press.

A balanced and straightforward presentation of the underlying mathematics of statistical analysis of meteorological data.

References

Abbas, M. A. & Latham, J. (1969). The electrofreezing of supercooled water drops. *J. Meteorol. Soc. Japan*, **47**, 65–74.

Allan, R. J. (2000). ENSO and climatic variability in the past 150 years. In *El Niño and the Southern Oscillation*, ed. H. F. Diaz & V. Markgraf, pp. 3–55. Cambridge: Cambridge University Press.

Alley, R. B., *et al.* (1993). Abrupt increase in Greenland snow accumulation at the end of the Younger Dryas event. *Nature*, **362**, 527–9.

Ambaum, M. P. H., Hoskins, B. J. & Stephenson, D. B. (2001). Arctic Oscillation or North Atlantic Oscillation? *J. Climate*, **14**, 3495–507.

Ammann, C. M. & Naveau, P. (2003). Statistical analysis of tropical explosive volcanism occurrences over the last 6 centuries. *Geophys. Res. Lett.*, **30**, No. 5, 1210, doi: 10.1028/2002GLO16388.

Appenzeller, C., Stocker T. F. & Anklin, M. (1998). North Atlantic Oscillation dynamics recorded in Greenland ice cores. *Science*, **282**, 446–9.

Baillie, M. G. L. (1995). *A Slice Through Time: Dendrochronology and Precision Dating*. London: Batsford.

Baldwin, M. P., *et al.* (2001). The quasi-biennial oscillation. *Rev. Geophys.*, **32**, 179–229.

Baldwin, M. P., *et al.* (2003). Stratospheric memory and skill of extended-range weather forecasts. *Science*, **301**, 636–40.

Baliunas, S., Frick, P., Sokoloff, D. & Soon, W. (1997). Time scales and trends in the Central England Temperature data (1659–1990): a wavelet analysis. *Geophys. Res., Lett.*, **24**, 1351–4.

Barnola, J. M., *et al.* (1987). Vostok ice core provides 160 000 year record of atmospheric CO_2. *Nature*, **329**, 410–16.

Barnston, A. G. & Livezey, R. E. (1989). A closer look at the effect of the 11-year solar cycle and the quasi-biennial oscillation on Northern Hemisphere

700 mb height and extratropical North American surface temperature. *J. Climate*, **2**, 1295–313.

Bath, M. (1974). *Spectral Analysis in Geophysics*. Amsterdam: Elsevier Scientific Publishing Co.

Battisti, D. S., & Hirst, A. C. (1989). Interannual variability in the tropical atmosphere–ocean system: influence of the basic state and ocean geometry. *J. Atmos. Sci.* **46**, 1687–712.

Behera, S. K. & Yamagata, T. (2001). Subtropical SST dipole events in the southern Indian Ocean. *Geophys. Res. Lett.*, **28**, 327–30.

Benzi, R., Parisi, G., Sutera, A. & Vulpiani, A. (1982). Stochastic resonance in climatic change. *Tellus*, **34**, 10–16.

Berger, A. (1990). Relevance of medieval Egyptian and American dates to the study of climatic and radiocarbon variability. In *The Earth's Climate and Variability of the Sun over Recent Millennia*, ed. J. C. Pecker and S. K. Runcorn, pp. 119–29. London: Royal Society.

Berger, A. Melice, J. L. & van der Mersch, I. (1990). Evolutive spectral analysis of sunspot data over the past 300 years. In *The Earth's Climate and Variability of the Sun over Recent Millennia*, eds. J. C. Pecker and S. K. Runcorn, pp. 131–42. Royal Society, London.

Beveridge, W. H. (1921). Weather and harvest cycles. *Econ. J.*, **31**, 429–47.

Biondi, F., Gershunov, A. & Cayan, D. R. (2001). North Pacific decadal climate variability since AD 1661. *J. Climate*, **14**, 5–10.

Bjerknes, J. (1969). Atmospheric teleconnections from the equatorial Pacific. *Mon. Wea. Rev.*, **97**, 163–72.

Black, D. E., *et al.* (1999). Eight centuries of North Atlantic Ocean atmosphere variability. *Science*, **286**, 1709–13.

Blunier, T. & Brook, E. J. (2001). Timing of millennial-scale climate change in Antarctica and Greenland during the last glacial period. *Science*, **291**, 109–12.

Bond, G. C. & Lotti, R. (1995). Iceberg discharges into the North Atlantic on millennial time scales during the last deglaciation. *Science*, **267**, 1005–10.

Bond, G., *et al.* (1992). Evidence for massive discharges of icebergs into the North Atlantic Ocean during the last glacial period. *Nature*, **360**, 245–9.

Bond, G., *et al.* (1997). A pervasive millennial-scale cycle in North Atlantic Holocene and Glacial climates. *Science*, **278**, 1257–65.

Bond, G., *et al.* (2001). Persistent solar influence on North Atlantic climate during the Holocene. *Science*, **294**, 2130–6

Briffa, K. R., *et al.* (1990). A 1400-year tree-ring record of summer temperatures in Fennoscandia. *Nature*, **346**, 434–9.

(1995). Unusual twentieth-century summer warmth in a 1000-year temperature record from Siberia. *Nature*, **376**, 156–9.

Broecker, W. S. (1975). Climate change: are we on the brink of a pronounced global warming? *Science*, **189**, 460–3.

(1994). Massive iceberg discharges as triggers for global climate change. *Nature*, **372**, 421–5.

(1995). Chaotic climate. *Sci. Am.*, **267**, No. 11, 44–50.

(1997). Thermohaline circulation, the Achilles Heel of our climate system: will man-made CO_2 upset the current balance? *Science*, **278**, 1582–8.

(1998). Paleocean circulation during the last deglaciation: a bipolar seesaw? *Paleoceanography*, **13**, 119–21.

Burroughs, W. J. (1978). On running means and meteorological cycles. *Weather*, **33**, 101–9.

(1982). Why do cold Decembers in England come at the end of each century? *Weather*, **37**, 205–6.

(1997). *Does the Weather Really Matter?* Cambridge: Cambridge University Press.

Campbell, I. D., *et al.* (2000). Millennial-scale rhythms in peatlands in the western interior of Canada and in the global carbon cycle. *Quatern. Res.*, **54**, 321–7.

Cavalieri, D., Gloersen, P., Parkinson, D. L., Cosimo, J. C. & Zwally, H. J. (1997). Observed hemispheric symmetry in global sea ice changes. *Science*, **278**, 1104–6.

Chapman, M. J. & Shackleton, N. J. (2000). Evidence of 550-year and 1000-year cyclicities in North Atlantic circulation patterns during the Holocene. *The Holocene*, **10**, 287–91.

Clegg, S. L. & Wigley, T. M. L. (1984). Periodicities in precipitation in Northeast China. *Geophys. Res. Lett.*, **11**, 1219.

CLIMAP Project Members (1976). The surface of the ice-age Earth. *Science* **191**, 1131–7.

Cobb, K. M., Charles, C. D. & Hunter, D. E. (2001). A central tropical Pacific coral demonstrates Pacific, Indian and Atlantic decadal climate connections. *Geophys. Res., Lett.*, **28**, 2209–12.

Cobb, K. M., Charles, C. D., Cheng, H. & Edwards, R. L. (2003). El Niño/Southern Oscillation and tropical climate during the last millennium. *Nature*, **424**, 271–6.

Cohen, T. J. & Lintz, P. R. (1974). Long term periodicities in the sunspot cycle. *Nature*, **250**, 398–400.

Cohen, T. J. & Sweetser, E. I. (1975). The 'spectra' of the solar cycle data for Atlantic tropical cyclones. *Nature*, **256**, 295–6.

Cole, J. E., Dunbar, R. B., McClanahan, T. R. & Muthiga, N. A. (2000). Tropical Pacific forcing of decadal SST variability in the western Indian Ocean over the past two centuries. *Science*, **287**, 617–19.

Cook, E. R., Briffa, K. R., Meko, D. M., Graybill, D. A. & Funkhouser, G. (1995). The 'segment length curse' in long tree-ring chronology development for palaeoclimatic studies. *The Holocene*, **5**, 229–37.

Cook, E. R., Meko, D. M. & Stockton, C. W. (1997) A new assessment of possible solar and lunar forcing of the bidecadal drought rhythm in the western United States. *J. Climate*, **10**, 1343–56.

Craddock, J. M. (1968). *Statistics in the Computer Age.* London: English University Press.

Crowley, K. D., Duchan, C. E. & Rhi, J. (1986). Climate record in varved sediments in Eocene Green River formation. *J. Geophys. Res.*, **91**, 8637–48.

Currie, R. G. (1981). Evidence of 18.6 year M_N signal in temperature and drought conditions in North America since AD 1800. *J. Geophys. Res.* **86**, 11 055–64.

(1987). In *Climate History, Periodicity and Predictability*, ed. M. R. Rampino, J. E. Sanders, W. S. Newman, & L. K. Konigsson, New York: Van Nostrand Reinhold.

(1993). Luni-solar 18.6- and solar cycle 10- to 11-year signals in USA air temperature records. *Int. J. Climatol.*, **13**, 31–50.

Currie, R. G. & O'Brien, D. P. (1988). Periodic 18.6 year and cyclic 10- to 11-year signals in the Northest United States precipitation data. *Int. J. Climatol.*, **8**, 255–81.

Czaja, A. & Frankignoul, C. (2002). Observed impact of Atlantic SST anomalies on the North Atlantic Oscillation. *J. Climate*, **15**, 31–50.

Dansgaard, W. & Oeschger, H. (1989). In *The Environmental Record in Glaciers and Ice Sheets*, ed. H. Oeschger, H. & Langway, C. C., pp. 287–318. Chichester, UK: Wiley.

Dansgaard, W., Johnsen, S. J., Clausen, H. B. & Langway, C. C. (1973). Climatic record revealed by Camp Century ice core. In *The Late Cenozoic Ice Ages*, ed. K. K. Turekian, pp. 37–56. Yale: Yale University Press.

Dansgaard, W., *et al.* (1993). Evidence of general instability of past climate from a 250-kyr ice-core record. *Nature*, **364**, 218–20.

De Geer, G. (1929). Solar registration by pre-Quaternary varve-shales. *Geogr. Ann.* **11**, 242–6.

de la Mare, W. K. (1997). Abrupt mid-twentieth-century decline in Antarctic sea-ice extent from whaling records. *Nature*, **389**, 57–60.

Delworth, T. L. (1996). North Atlantic interannual variability in a coupled ocean–atmosphere model. *J. Climate*, **9**, 2356–75.

Delworth, T. L. & Mann, M. E. (2000). Observed and simulated multidecadal variability in the Northern Hemisphere. *Clim. Dynam.*, **16**, 661–76.

Delworth, T. L., Manabe, S. & Stouffer, R. J. (1997). Multidecadal climate variability in the Greenland Sea and surrounding regions: a coupled model simulation. *Geophys. Res. Lett.*, **24**, 257–60.

Denton, G. H. & Karlén, W. (1973). Holocene climate variations – their pattern and possible cause. *Quatern. Res.*, **3**, 155–205.

Diaz, H. F. & Markgraf, V. (eds.) (2000). *El Niño and the Southern Oscillation.* Cambridge: Cambridge University Press.

Diaz, H. F., Hoerling, M. P. & Eischeid, J. K. (2001). ENSO variability: teleconnections and climate change. *Int. J. Climatol.*, **21**, 1845–62.

Dombros, M. & Gongbing, P. (1988). *The Climate of China.* New York, Heidelberg: Springer-Verlag.

Dommenget, D. & Latif, M. (1999). Interannual to decadal variability in the tropical Atlantic. *J. Climate*, **13**, 777–92.

Douglass, A. E. (1919). *Climate Cycles and Tree Growth.* Washington, DC: Carnegie Institute of Washington.

Dunbar, R. B., Wellington, G. M., Colgan, M. W. & Glynn, P. W. (1994). Eastern Pacific sea-surface temperature since 1600 AD from the delta ^{18}O record of climate variability in Galapagos corals. *Paleoceanography*, **9**, 291–315.

Dyer, T. G. J. (1978). Persistence and monthly mean temperatures over Central England. *Weather*, **33**, 141–8.

Eddy, J. A. (1976). The Maunder minimum. *Science*, **192**, 1189–202.

Eddy, J. A., Gilliland, R. L. & Hoyt, D. V. (1982). Changes in the solar constant and climatic effects. *Nature*, **300**, 689–93.

Eden, C. & Willebrand, J. (2001). Mechanism of interannual to decadal variability of the North Atlantic Circulation. *J. Climate*, **14**, 2266–2280.

Egbert, G. D. & Ray, R. D. (2000). Significant dissipation of tidal energy in the deep ocean inferred from satellite altimeter data. *Nature*, **405**, 775–8.

Eichkorn, S. Wilhelm, S. Aufmhoff, H. Wohlfrom, K. H. & Arnold, F. (2002). Cosmic ray-induced aerosol-formation: First observational evidence from aircraft-based ion mass spectrometer measurements in the upper troposphere. *Geophys. Res. Lett.*, **29**, 10.1029/2002GLO15044.

Elsner, J. B. & Tsonis, A. A. (1991). Do bidecadal oscillations exist in the global temperature record? *Nature*, **353**, 551–3.

Eltahir, E. A. B. & Wang, G. (1999). Nilometers, El Niño, and climate variability. *Geophys. Res. Lett.* **26**, 489–92.

Elton, C. S. (1924). Periodic fluctuations in the number of animals: their causes and effects. *Br. J. Exp. Biol.* **2**, 119–63.

Emiliani, C. (1955). Pleistocene temperatures. *J. Geology*, **63**, 538–78.

Enfield, D. B., Mestas-Nuñez, A. M. & Trimble, P. J. (2001). The Atlantic multidecadal oscillation and its relation to rainfall and river flows in the continental U.S. *Geophys. Res. Lett.*, **28**, 2077–80.

Esper, J. Cook, E. R. & Schweingruber, F. H. (2002). Low frequency signals in long tree ring chronologies for reconstructing past temperature variability. *Science*, **295**, 2250–3.

Evans, M. N., *et al.* (2001). Support for tropically-driven Pacific decadal variability based on paleoproxy evidence. *Geophys. Res. Lett.*, **28**, 3689–92.

Federov, A. V. *et al.* (2003). How predictable is El Niño? *Bull. Am. Meteorol. Soc.*, **84**, 911–19.

Folland, C. K. & Salinger, M. J. (1997). Surface temperature trends and variations in New Zealand and the surrounding ocean. *Int. J. Climatol.*, **15**, 1195–218.

Foukal, P. & Lean, J. (1990). An empirical model of total solar irradiance variation between 1874 and 1986. *Science*, **247**, 556–8.

Gagan, M. K., *et al.* (1998). Temperature and surface-ocean water balance of the mid-Holocene tropical western Pacific. *Science*, **279**, 1014–18.

Gallego, G. & Cressi, P. (2001). Decadal variability of two oceans and an atmosphere. *J. Climate*, **14**, 2815–32.

Ganopolski, A. & Rahmstorf, S. (2002). Abrupt glacial climate changes due to stochastic resonance. *Phys. Rev. Lett.*, **88**, 038501-1–4.

Giovanelli, R. (1984). *Secrets of the Sun.* Cambridge: Cambridge University Press.

Gleissberg, W. (1958). The eighty-year sunspot cycle. *J. Br. Astron. Assoc*, **68**, 148–52.

Gong, D. & Wang, S. (1999). Definition of Antarctic Oscillation Index. *Geophys. Res. Lett.*, **26**, 459–62.

Gordon, A. H. (1976). The frequency distribution of changes in mean temperature from one month to the next. *Weather*, **31**, 197–200.

Graham, N. E. & White, W. B. (1988). The El Niño cycle: a natural oscillator of the Pacific Ocean–atmosphere system. *Science*, **240**, 1293–302.

Gray, W. M. (1990). Strong association between West African rainfall and US landfall of intense hurricanes. *Science*, **249**, 1251–6.

Greenland Ice Core Project (GRIP) Members (1993). Climate instability during the last interglacial period recorded in the GRIP ice core. *Nature*, **364**, 203–7.

Gribbin, J. (1982). Stand by for bad winters. *New Sci.*, 28 October, 220–3.

Grootes, P. M. & Stuiver, M. (1997). Oxygen 18/16 variability in Greenland snow and ice with 10^{-3} to 10^5 year time resolution. *J. Geophys. Res.*, **102**, 26 455–67.

Grootes, P. M., Stuiver, M., White, J. W. C., Johnsen, S. & Jouzel, J. (1993). Comparison of oxygen isotope records from the GISP 2 and GRIP Greenland ice cores. *Nature*, **366**, 552–4.

Grove, J. M. (1988). *The Little Ice Age.* Methuen.

Haigh, J. D. (1999). A GCM study of climate change in response to the 11-year solar cycle. *Q. J. R. Meteorol. Soc.*, **125**, 871–92.

(2000). Solar variability and climate. *Weather*, **55**, 399–405.

Hameed, S., *et al.* (1983). An analysis of periodicities in the 1470 to 1979 Beijing precipitation record. *Geophys. R. Lett.*, **10**, 436–9.

Harrison, G. (2002). Twentieth century secular decrease in the atmospheric potential gradient. *Geophys. Res. Lett.*, **29**, 10.1029/2002GL014878.

Hastenrath, S. (2002). Dipoles, temperature gradients, and tropical climate anomalies. *Bull. Am. Meteorol. Soc.*, **83**, 735–8.

Hastenrath, S. & Heller, L. (1977). Dynamics of climatic hazards in Northeast Brazil. *Q. J. R. Meteorol. Soc.*, **103**, 77–92.

Hays, J. D., Imbrie, J. & Shackleton, N. J. (1976). Variations in the Earth's orbit: pacemaker of the Ice Ages. *Science*, **194**, 1121–32.

Heinrich, H. (1988). Origin and consequences of cyclic ice rafting in the northeast Atlantic Ocean during the past 130,000 years. *Quatern. Res.*, **29**, 142–52.

Henderson-Sellers, A. (1992). Continental cloudiness changes this century. *GeoJournal*, **27**, 255–62.

Hendon, H. H., Liebmann, B. & Glick, J. D. (1998). Oceanic Kelvin waves and the Madden–Julian oscillation. *J. Atmos. Sci.*, **55**, 88–100.

Herbert, T. D., *et al.* (2001). Collapse of the California current during glacial maxima linked to climate change on land. *Science*, **293**, 71–6.

Hibler, W. D. III & Johnson, S. J. (1972). The 20-year cycle in Greenland ice core records. *Nature*, **280**, 429–34.

Holton, J. R. & Lindzen R. S. (1972). An up-dated theory of the equatorial quasi-biennial oscillation in the tropical stratosphere. *J. Atmos. Sci.*, **29**, 1076–1080.

Hurrell, J. W. (1995). Decadal trends in the North Atlantic Oscillation: regional temperatures and precipitation. *Science*, **269**, 676–8.

(1996). Influences of variations in extratropical wintertime teleconnections on Northern Hemisphere temperature. *Geophys. Res. Lett.* **23**, 665–8.

Ichi-Kuma, K. (1990). A QBO in the intensity of the intraseasonal oscillation. *Int. J. Climatol.*, **10**, 263–78.

Imbrie, J. & Imbrie, J. Z. (1980). Modelling the climatic response of orbital variations. *Science*, **207**, 943–53.

Imbrie, J., *et al.* (1992). On the structure and origin of major glaciation cycles. 1. Linear responses to Milankovitch forcing. *Paleoceanography*, **7**, 701–38.

(1993a). On the structure and origin of major glaciation cycles. 2. The 100,000 year cycle. *Paleoceanography*, **8**, 699–735.

Imbrie, J., Mix, A. C. & Martinson, D. G. (1993b). Milankovitch theory viewed from Devil's Hole. *Nature*, **363**, 531–3.

IPCC (1990). *Climate Change: The IPCC Scientific Assessment*, ed. J. T. Houghton, G. J. Jenkins & G. G. Ephraums. Cambridge: Cambridge University Press.

(1992). *Climate Change 1992: The Supplementary Report to IPCC Scientific Assessment*, ed., J. T. Houghton, B. A. Callander & S. K. Varney. Cambridge: Cambridge University Press.

(1994). *Climate Change 1994: Radiative Forcing of Climate and an Evaluation of the IPCC IS92 Emission Scenarios*, ed. J. T. Houghton, L. G. Meira Filho,

J. Bruce, Hoesung Lee, B. A. Callendar, E. Haites, N. Harris, & K. Maskell. Cambridge: Cambridge University Press.

(1995). *Climate Change 1995: The Science of Climate Change*, ed. J. T. Houghton, L. G. Meira Filho, B. A. Callendar, N. Harris, A. Kattenberg & K. Maskell. Cambridge: Cambridge University Press.

(2001). *Climate Change 2001: The Scientific Basis*, ed. J. T. Houghton, Y. Ding, D. Griggs, M. Noguer, P. J. van der Linden, X. Dai, K. Maskell & C. A Johnson. Cambridge: Cambridge University Press.

Isdale, P. J., Stewart, B. J., Tickle K. S. & Lough, J. M. (1998). Palaeohydro-logical variation in a tropical river catchment: a reconstruction using fluorescent bands in corals of the Great Barrier Reef, Australia. *The Holocene*, **8**, 1–8.

James, I. N. & James, P. N. (1989). Ultra-low frequency variability in a simple atmospheric circulation model. *Nature*, **342**, 53–5.

Jiang, N. Neelin, J. D. & Ghil, M. (1995). Quasi-quadrennial and quasi-biennial variability in the equatorial Pacific. *Climate Dynam.*, **12**, 101–12.

Kalnay, E., *et al.* (1996). The NCEP/NCAR 40-year reanalysis project. *Bull. Am. Meteorol. Soc.*, **77**, 437–71.

Kerr, R. A. (1988). Sunspot-weather link holding up. *Science*, **242**, 1124.

Kiehl, J. T. & Trenberth, K. E. (1997). Earth's annual global mean energy budget. *Bull. Am. Meteorol. Soc.*, **78**, 197–208.

Kniveton, D. R. & Todd, M. C. (2001). On the relationship of cosmic ray flux and precipitation. *Geophys. Res. Lett.*, **28**, 1527–30.

Kwok, R. & Comiso, J. (2002). Southern ocean climate and sea ice anomalies associated with the southern oscillation. *J. Climate*, **15**, 487–501.

Labitzke, K. (2001). The global signal of the 11-year sunspot cycle in the strato-sphere: differences between solar maxima and solar minima. *Meteorolo-gische Zeitschrift*, **10**, 901–8.

Labitzke, K. & van Loon, H. (1990). Association between the 11-year solar cycle, the quasi-biennial oscillation and the atmosphere: a summary of recent work. In *The Earth's Climate and Variability of the Sun over Recent Millen-nia*, ed. J. C. Pecker and S. K. Runcorn, pp. 179–82. London: Royal Society.

(1999). *The Stratosphere: Phenomena, History and Relevance*. New York: Springer-Verlag.

Lamb, H. H. (1972). *Climate: Present, Past and Future*. Volume 1. London: Methuen.

Lamb, P. J. (1978). Large-scale tropical Atlantic circulation patterns associated with Subsaharan weather anomalies. *Tellus*, **30**, 240–1.

Lambert, D. (1988). *The Cambridge Guide to the Earth*. Cambridge: Cambridge University Press.

Landsberg, H. E., *et al.* (1963). Surface signs of the biennial atmospheric pulse. *Mon. Wea. Rev.*, **91**, 549–56.

Landsea, C. W., Gray, W. M., Mielke Jr, P. W. & Berry, J. K. (1994). Seasonal fore-casting of Atlantic hurricane activity. *Weather*, **49**, 273–84.

Latif, M. (2001). Tropical Pacific/Atlantic Ocean interactions at multi-decadal time scales. *Geophys. Res. Lett.*, **28**, 539–42.

Lau, K.-M., Kim, K-M. & Shen, S. S. P. (2002). Potential predictability of seasonal precipitation over the United States from canonical ensemble correlation predictions. *Geophys. Res. Lett.*, **29**, 1–4.

Lean, J. L. (2000). Evolution of the sun's spectral irradiance since the Maunder minimum. *Geophys. Res. Lett.*, **27**, 2425–28.

Lean, J. L. & Rind, D. (2001). Earth's response to a variable sun. *Science*, **292**, 234–6.

Lean, J. L., White, O. R., Livingston, W. C. & Picone, J. M. (2001). Variability of a composite chromospheric irradiance index during the 11-year activity cycle and over longer time periods. *J. Geophys. Res.*, **106**, 10645–58.

Lejenas, H. (1995). Long term variations of atmospheric blocking in the northern hemisphere. *J. Meteorol. Soc. Japan*, **73**, 79–89.

Lejenas, H., & Okland, H. (1983) Characteristics of Northern Hemisphere block-ing as determined from a long time series of observational data. *Tellus*, **35A**, 350–62.

Le Roy Ladurie, E. & Baulant, M. (1980). Grape harvests from the fifteenth through the nineteenth centuries. *J. Interdisciplin. Hist.*, **10**, 839–49.

Lindstrom, J. (1997). Solar activity and hare dynamics: a cross-continental com-parison. *Am. Natural.*, **149**, 765–75.

Lindzen, R. S. (1987). On the development of the theory of the QBO. *Bull. Am. Meteorol. Soc*, **68**, 329–37.

Linsley, B. K., Wellington, G. M. & Schrag, D. P. (2000). Decadal sea surface tem-perature variability in the subtropical South Pacific from 1726 to 1997 AD. *Science*, **290**, 1145–8.

Liu H. S. & Chao B. F. (1998). Wavelet spectral analysis of the Earth's orbital vari-ations and palaeoclimatic cycles. *J. Atmos. Sci.*, **55**, 227–36.

Lockwood, M. & Stamper, R. (1999). Long term drift in the coronal source mag-netic flux and total solar irradiance. *Geophys. Res. Lett.*, **26**, 2461–5.

Lockwood, M., Stamper, R. & Wild, M. N. (1999). A doubling of the Sun's coronal magnetic field during the last 100 years. *Nature*, **399**, 437–9.

Lorenz, E. N. (1963). Deterministic nonperiodic flow. *J. Atmos. Sci.*, **20**, 130–41.

Lu, H. *et al.* (2000) Variability of East Asian winter monsoon in Quaternary climatic extremes in North China. *Quatern. Res.*, **54**, 321–7.

Lundin, R., Eliasson, L. & Murphree, J. S. (1991). The quiet time aurora. In *Auroral Physics*, ed. C.-I. Meng, M. J. Rycroft & L. A. Frank. Cambridge: Cambridge University Press.

Luterbacher, J. *et al.* (2002). Extending the North Atlantic Oscillation recon-structions back to 1500. *Atmos. Sci. Lett.*, **2**, 114–24.

Madden, R. A. & Julian, P. R. (1971). Detection of a 40–50 day oscillation in the zonal wind in the tropical Pacific. *J. Atmos. Sci.*, **28**, 702–8.

(1972). Description of global-scale circulation cells in the tropics with a 40–50 day period. *J. Atmos. Sci.*, **29**, 1109–23.

(1994). Observations of the 40–50 day tropical oscillation: a review. *Mon. Wea. Rev.*, **122**, 814–37.

Manabe, S. & Stouffer, R. J. (1988). Two stable equilibria of a coupled ocean-atmosphere model. *J. Climate*, **1**, 841–66.

Manley, G. (1974). Central England temperatures: monthly means 1659 to 1973. *Q. J. R. Meteorol. Soc.*, **100**, 389–405.

Mantua, N. J., Hare, S. R., Zhang, Y., Wallace, J. M., & Francis, R. C. (1997). A Pacific interdecadal climate oscillation with impacts on salmon production. *Bull. Am. Meteorol. Soc.*, **78**, 1069–79.

Marcus, P. S., Sommeria, J., Meyers, S. D. & Swinney, H. L. (1990). Models of the Great Red Spot, *Nature*, **343**, 517–18.

Markson, R. (1978). Solar modification of atmospheric electrification and possible implications for the Sun–weather relationship. *Nature*, **244**, 197–200.

Marshall, J., *et al.* (2001). North Atlantic climate variability: phenomena, impacts and mechanisms, *Int. J. Climatol.*, **21**, 1863–98.

Martinson, D. G., *et al.* (1987). Age dating and the orbital theory of ice ages: development of a high-resolution 0 to 300,000 years chronostratigraphy. *Quatern. Res.*, **27**, 1–29.

Mason, B. J. (1976). Towards the understanding and prediction of climatic variations. *Q. J. R. Meteorol. Soc.*, **102**, 478–98.

Maunder, E. W. (1922). The prolonged sunspot minimum 1645–1715. *J. Br. Astron. Assoc.*, **32**, 140–5.

May, B. R. & Hitch, T. J. (1989). Periodic variations in extreme hourly rainfall in the UK. *Meteorol. Mag.*, **118**, 45–50.

McDermott, F., Mattey, D. P. & Hawkesworth, C. (2001). Centennial-scale Holocene climate variability revealed by a high-resolution speleothem $\delta^{18}O$ Record from SW Ireland. *Science*, **294**, 1328–31.

McPhaden, M. J., Delcroix, T., Hanawa, K., Kuroda, Y., Meyers, G. Picaut, J. & Swenson, M. (2001). The El Niño/Southern Oscillation (ENSO) observing system. In *Observing the Ocean in the 21st Century*, pp. 231–46. Melbourne, Australia: Australian Bureau of Meteorology.

Mitchell, J. M. (1990) Climatic variability: past, present & future. *Climatic Change*, **16**, 231–46.

Mitchell, J. M., Stockton, C. W. & Meko, D. M. (1979). Evidence of a 22-year rhythm of drought in the Western United States related to the Hale Solar Cycle since the 17th century. In *Solar–Terrestrial Influences on Weather and Climate*, ed. B. M. McCormac & T. A. Seliga, pp. 125–43. D. Reidel Publishing Co.

Mitton, S. M. (ed.) (1977). *Cambridge Encyclopedia of Astronomy.* Cambridge: Cambridge University Press.

Mock, S. J. & Hibler, W. D. III (1976). The 20-year oscillation in American temperature records. *Nature,* **261,** 484–6.

Montgomery, R. B. (1940). Report on the work of G. T. Walker. *Mon. Wea. Rev., Supp. No. 39,* 1–22.

Muller, R. A. & MacDonald, G. J. (1997). Glacial cycles and astronomical forcing. *Science,* **277,** 215–18.

Namias, J. (1985). Some empirical evidence of influence of snow cover on temperature and precipitation. *Mon. Wea. Rev.,* **113,** 1542–53.

Neff, U., *et al.* (2001). Strong coherence between solar variability and the monsoon in Oman between 9 and 6 kyr ago. *Nature,* **411,** 290–3.

Neftel, A., Oeschger, H. & Suess, H. E. (1981). Secular non-random variations of cosmogenic carbon-14 in the terrestrial atmosphere. *Earth Planet. Sci. Lett.,* **56,** 127–47.

Newell, N. E., Newell, R. E., Hsuing, J. & Wu, Z. (1989). Global marine temperature variation and the solar magnetic cycle. *Geophys. Res. Lett.,* **16,** 311–14.

Nicholas, F. J. & Glasspoole, J. (1932). General monthly rainfall for England and Wales, 1727 to 1931. *Br. Rainfall,* **1931,** 299–306.

Nobre, P. & Shukla, J. (1996). Variations of sea surface temperature, wind stress, and rainfall over the tropical Atlantic and South America. *J. Climate,* 9, 2464–79.

Oeschger, H. & Beer, J. (1990). In the past 5000 years history of solar modulation of cosmic radiation from ^{10}Be and ^{14}C studies. *The Earth's Climate and Variability of the Sun over Recent Millennia,* ed. J. C. Pecker and S. K. Runcorn, pp. 73–82. London: Royal Society.

Oix *et al.* (1999). Spectral analysis of a 1000-year stalagmite lamina-thickness record from Shihua Cavern, Beijing, China, and its climatic significance. *The Holocene,* **9,** 689–94.

Okal, E. & Anderson, D. L. (1975). On the planetary theory of sunspots. *Nature,* **253,** 511–13.

Oliver, R., Ballester, J. L. & Baudin, F. (1998). Emergence of magnetic flux on the Sun as the cause of a 158-day periodicity in sunspot areas. *Nature,* **394,** 552–3.

Paillard, D. (1998). The timing of Pleistocene glaciations from a simple multiple-state model. *Nature,* **391,** 378–81.

Palmer, T. (1993). A nonlinear dynamical perspective on climate change. *Weather,* **48,** 314–25.

Parker, D. E., Legg, T. P. & Folland, C. K. (1992). A new daily Central England temperature series, 1772–1991. *Int. J. Climatol.* **12,** 317–42.

Pestiaux, P., *et al.* (1988). Paleoclimatic variability at frequencies ranging from one cycle per 20 kyr to one cycle per kyr: evidence of non-linear behaviour of the climate system. *Climate Change,* **12,** 9–37.

Peterson, R. G. & White, W. (1998). Slow oceanic teleconnections linking the Antarctic Circumpolar Wave with tropical ENSO. *J. Geophys. Res.*, **103**, 24 573–83.

Petit, J. R., *et al.* (1999). Climate and atmospheric history of the past 420,000 years from the Vostok ice core, Antarctica. *Nature*, **399**, 429–36.

Philander, S. G. H. (1983). El Niño Southern Oscillation. *Nature*, **302**, 295–301.

(1990). *El Niño, La Niña, and the Southern Oscillation*. New York: Academic Press.

Pittock, A. B. (1978). A critical look at long term sun–weather relationships. *Rev. Geophys. Space Phys.*, **16**, 400–20.

(1983). Solar variability, weather and climate: an update. *Q. J. R. Meteorol. Soc.*, **109**, 23–55.

Plaut, G., Ghil, M. & Vautard, R. (1995). Interannual and interdecadal variability in 335 years of Central England temperatures. *Science*, **250**, 324–7.

Pool, R. (1989). Ecologists flirt with chaos. *Science*, **243**, 310–13.

Rahmstorf, S. (2003). Timing of abrupt climate change: A precise clock. *Geophys. Res. Lett.*, **30**, No. 10, 1510, doi 10.1029/2003GL017115.

Rahmstorf, S. & Alley, R. (2002). Stochastic resonance in glacial climate. *Eos*, **83**, 129–35.

Raisbeck, G. M., Yiou, F., Jouzel, J & Petit, J. R. (1990). ^{10}Be and δH in polar ice cores as a probe of the solar variability's influence on climate. In *The Earth's Climate and Variability of the Sun over Recent Millennia*, eds. J. C. Pecker and S. K. Runcorn, pp. 65–72. London: Royal Society.

Rajagopalan, B., Kushnir, Y. & Torre, Y. M. (1998). Observed decadal midlatitude and tropical atlantic variability. *Geophys. Res. Lett.*, **25**, 3967–70.

Ram, M. & Stolz, M. (1999). Possible solar influences on the dust profile of the GISP2 ice core from central Greenland. *Geophys. Res. Lett.*, **26**, 1043–6.

Ram, M. Stolz, M. & Koenig, G. (1997). Eleven-year cycle of dust concentration variability observed in the dust profile of the GISP2 ice core from central Greenland: possible solar cycle connection. *Geophys. Res. Lett.*, **24**, 2359–62.

Ramanathan, R., *et al.* (1989). Cloud-radiative forcing and climate: results of the Earth radiation budget experiment. *Science*, **243**, 57–63.

Rampino, M. R., Sanders, J. E., Newman, W. S. & Konigson, L. K. (1987). *Climate: History, Periodicity and Predictability*. New York: Van Nostrand Reinhold.

Rex, D. F. (1950). Blocking action in the middle troposphere and its effects on regional climate. *Tellus*, **2**, 196–211 (Part I), 275–301 (Part II).

Rind, D. (2001). The Sun's role in climatic variations. *Science*, **296**, 673–7.

Roberts, W. O. & Olson, R. H. (1973). Geomagnetic storms and wintertime 300 mb trough development in the North Pacific–North America area. *J. Atmos. Sci.*, **30**, 135–40.

Robertson, A. W., Mechoso, C. R. & Kim, Y.-J. (2000). The influence of Atlantic sea surface temperature anomalies on the North Atlantic oscillation. *J. Climate*, **13**, 122–38.

Rodwell, M. J. Rowell, D. P. & Folland, C. K. (1999). Oceanic forcing of the wintertime North Atlantic Oscillation and European climate. *Nature*, **398**, 320–3.

Roosen, R. G., Harrington, R. S., Giles, J. & Browning, I. (1976). Earth tides, volcanoes and climate change. *Nature*, **261**, 680–2.

Ropelewski, C. F. & Halpert, M. S. (1987). Global and regional scale preciptation patterns associated with the El Niño Southern Oscillation. *Mon. Wea. Rev.*, **115**, 1606–26.

Ropelewski, C. F., Halpert, M. S. & Wang, X. (1992). Observed tropical biennial variability and its relationship to the Southern Oscillation. *J. Climate*, **5**, 536–47.

Rossow, W. B. & Schiffer, R. A. (1999). Advances in understanding clouds from ISCCP. *Bull. Am. Meteorol. Soc.*, **80**, 2261–87.

Ruiz-Barradas, A., Carton, J. A. & Nigam, S. (2000). Structure of interannual-to-decadal climate variability in the tropical Atlantic sector. *J. Climate*, **13**, 3285–97.

Saito, K. & Cohen, J. (2003). The potential role of snow cover in forcing interannual variability of the major Northern Hemisphere mode. *Geophys. Res. Lett.*, **30**, No. 6, 1302, doi: 10.1029/2002GL016341.

Saji, N. J., Goswami, B. N., Vinayachandran, P. N. & Yamagata, T. (1999). A dipole mode in the tropical Indian Ocean. *Nature*, **401**, 360–3.

Schlesinger, M. E. & Ramankutty, N. (1994). An oscillation in the global climate system of period 65–70 years. *Nature*, **367**, 723–6.

Schmutz, C., *et al.* (2000). Can we trust proxy-based NAO index reconstructions? *Geophys. R. Lett.*, **27**, 1135–8.

Schonweise, C. D. (1980). Statistical comparison of central England annual and monthly air temperature variability. *Meteorol. Mag.* **109**, 101–12.

Schwabe, H. (1844). Solar observations during 1843. *Astr. Nachr.*, **21**, 233–48.

Shackleton, N. J. & Opdyke, N. D. (1973). *Quatern. Res.*, **3**, 39.

Shaw, N. (1926–1932). *Manual of Meteorology.* Cambridge: Cambridge University Press.

   (1933). *The Drama of the Weather.* Cambridge: Cambridge University Press.

Selley, R. C. (1988). *Applied Sedimentology.* London: Academic Press.

Sherratt, A. (1980). *Cambridge Encyclopedia of Earth Sciences.* Cambridge: Cambridge University Press.

Shindell, D. T., Rind, D., Balachandran, N., Lean, J. & Lonergan, P. (1999). Solar cycle variability, ozone and climate. *Science*, **284**, 305–8.

Slingo, J. M., *et al.* (1996). Intra seasonal oscillations in 15 atmospheric general circulation models: results from an AMIP diagnostic subproject. *Clim. Dyn.*, **12**, 325–57.

Slingo, J. M., Rowell, D. P. Sperber, K. R. & Nortley, F. (1999). On the predictability of the interannual behaviour of the Madden–Julian Oscillation and its relationship with El Niño. *Q. J. R. Meteorol. Soc.*, **125**, 583–609.

Smith, D. G. (1982). *Cambridge Encyclopedia of Earth Sciences*. Cambridge: Cambridge University Press.

Sommeria, J., Meyers S. D. & Swinney, H. L. (1988). A laboratory simulation of Jupiter's Great Red Spot. *Nature*, **331**, 374–6.

Soon, W. H., Posmentier, E. & Baliunas, S. L. (2000a). Climate hypersensitivity to solar forcing? *Ann. Geophys.*, **18**, 583–8.

Soon, W. H., Baliunas, S. L., Posmentier, E. & Okeke, P. (2000b). Variations in solar coronal hole area and terrestrial lower tropospheric temperature from 1979 to mid-1998: astronomical forcings of change in earth's climate? *New Astron.*, **4**, 563–79.

Steig, E. J. & Alley, R. B. (2002). Phase relationships between Antarctica and Greenland climate records. *Ann. Glaciol.*, **35**, 451–6.

Stommel, H. (1961). Thermohaline circulation with two stable regimes of flow. *Tellus*, **13**, 224–30.

Stone L., Saparin P. I., Huppert A., & Price C. (1998). El Niño chaos – the role of noise and stochastic resonance on the ENSO cycle. *Geophys. Res. Lett.*, **25**, 175–8.

Stuiver, M. & Braziunas, T. F. (1989). Atmospheric ^{14}C and century-scale solar oscillations. *Nature*, **338**, 405–8.

Stuiver, M. & Quay, P. D. (1980). Changes in atmospheric carbon-14 attributable to a variable Sun. *Science*, **207**, 11–19.

Stuiver, M., *et al.* (1998). INTCAL98 Radiocarbon age calibration, 24,000–0 cal BP. *Radiocarbon*, **40**, 1041–83.

Sturges, W. & B. G. Hong (2001). Gulf stream transport variability at periods of decades. *J. Phys. Oceanogr.*, **31**, No. 5, 1304–12.

Suarez, M. J., & Schopf, P. S. (1988). A delayed action oscillator for ENSO. *J. Atmos. Sci.* **45**, 3283–7.

Suess, H. E. & Linick, T. W. (1990). The C record in bristlecone pine wood of the past 8000 years based on the dendrochronology of the late C. W. Ferguson. *Phil. Trans. R. Soc. Lond.*, **A 330**, 403–412. (Also published in Pecker, & Runcorn (1990).)

Sutton, R. T. & Allen, M. R. (1997). Decadal predictability of the North Atlantic sea surface temperature and climate. *Nature*, **388**, 563–7.

Svensmark, H. and Friis-Christensen, E. (1997). Variation of cosmic ray flux and global cloud coverage – a missing link in solar–climate relationships. *J. Atmos. Solar–Terrestrial Phys.*, **59**, 1225–32.

Tabony, R. C. (1979). A spectral filter analysis of long-period records in England and Wales. *Meteorol. Mag.* **108**, 97–119.

Takahashi, M. (1999). Simulation of the stratospheric quasi-biennial oscillation in a general circulation model. *Geophys. Res. Lett.*, **26**, 1307–10.

Taylor, K. C., *et al.* (1993). The 'flickering switch' of late Pleistocene climate change. *Nature*, **361**, 432–6.

Thejll, P. (2001). Decadal power in land air temperatures: is it statistically significant? *J. Geophys. Re*s., **106**, 31693–704.

Thompson, J. M. T. & Stewart, H. B. (1986). *Nonlinear Dynamics and Chaos*. New York: John Wiley.

Thompson, D. W. C., Baldwin, M. P. & Wallace, J. M. (2002). Stratospheric connection to Northern Hemisphere wintertime weather: implications for prediction. *J. Climate*, **15**, 1421–8.

Thompson, D. W. J. & Wallace, J. M. (2000). Annular modes in the extratropical circulation. Part I: Month-to-month variability. *J. Climate*, **13**, 409–12.

Thompson, D. W. J., Wallace, J. M. & Hegerl, G. C. (2000). Annual modes in the extratropical circulation Part II: Trends. *J. Climate*, **13**, 1018–36.

Tinsley, B. A. (1988). The solar cycle and the QBO influences on the latitude of storm tracks in the North Atlantic. *Geophys. Res. Lett.*, **15**, 1000–16.

(1996). Correlations of atmospheric dynamics with solar wind-induced changes of air–earth current density into cloud top. *J. Geophys. Re*s., **101**, 29701–14.

Tinsley, B. A. & Yu, F. (2002). Effects of particle flux variations on clouds and climate. AGU Monograph *Solar Variability and its Effect on the Earth's Atmospheric and Climate System*, ed. C. Frolich, H. Hudson, J. Kuhn, J. McCormack, J. North, J. Pap, W. Sprigg, and S. T. Wu.

Torrence, C. & Compo, G. P. (1998). A practical guide to wavelet analysis. *Bull. Am. Meteorol. Soc.*, **79**, 61–78.

Tourre, Y. M., B. Rajagopalan, B., Kushnir, Y., Barlow, M. & White, W. B. (2001). Patterns of coherent decadal and interdecadal climate signals in the Pacific basin during the 20th century. *Geophy. Res. Lett.*, **28**, 2069–72.

Trenberth, K. E. (ed.) (1992). *Climate System Modelling*. Cambridge: Cambridge University Press.

Tripathi, S. N. & Harrison, R. G. (2002). Enhancement of contact nucleation by scavenging of charged aerosol particles. *Atmos. Res.*, **62**, 57–70.

Tsien, H. S. (1954). *Engineering Cybernetics*. New York: McGraw Hill.

Tyson, P. D. (1986). *Climate Change and Variability in Southern Africa*. Oxford: Oxford University Press.

Ulrych, T. J. & Bishop, T. N. (1975). Maximum entropy spectral analysis and autoregressive decomposition. *Rev. Geophys. Space Phys.* **13**, 183–200.

van Loon, H. & Labitzke, K. (2000). The influence of the 11-year solar cycle on the stratosphere below 30 km: a review. *Space Science Rev.*, **94**, 259–78.

van Loon, H. & Rogers, J. C. (1978). The seesaw in winter temperatures between Greenland and northern Europe. Part 1: General description. *Mon. Wea. Rev.*, **106**, 295–310.

Venegas, S. A. (2001). *Statistical Methods of Signal Detection in Climate*. Danish Center for Earth System Science, University of Copenhagen.

Vines, R. G. & Tomlinson, A. I. (1985). The Southern Oscillation and rainfall patterns in the Southern Hemisphere. *S. Afr. J. Sci.* **85**, 151–6.

Walker, G. T. (1927). World weather III, *Mem. R. Meteorol. Soc.*, **2**, No. 17. London.
    (1928). World Weather. *Q. J. R. Meteorol. Soc.*, **54**, 79–87.
    (1929). World Weather IV – some applications to seasonal foreshadowing. *Mem. R. Meteorol. Soc.*, **3**, No. 24. London.

Walker, G. T & Bliss, E. W. (1933). World weather V, *Mem. R. Meteorol. Soc.*, **4**, No. 36. London.

Wang Shao-Wu & Zhao Zong-ci (1981). Droughts and floods in China 1470–1979. In *Climate and History. Studies in Past Climates and their Impact on Man*, ed. T. M. L. Wrgley, M. J. Ingram & G. Farmer, pp. 271–88. Cambridge: Cambridge University Press.

Weaver, A. J., Sarachik, E. S. & Marotzke, J. (1992). Freshwater flux forcing of decadal and interdecadal oceanic variability. *Nature*, **353**, 836–8.

Webster, P. J., Moore, A. M. Loschnigg, J. P. & Leben, R. R. (1999). Coupled ocean–atmosphere dynamics in the Indian Ocean during 1997–98. *Nature*, **401**, 356–60.

White, W., & Peterson, R. G. (1996). An Antarctic circumpolar wave in surface pressure, wind, temperature, and sea ice extent. *Nature*, **380**, 699–702.

Whitlock, C. & Bartlein, P. J. (1997). Vegetation and climate change in northwestern America during the last 125 kyr. *Nature*, **388**, 57–61.

Wigley, T. M. L., Lough, J. M. & Jones, P. D. (1984). Spatial patterns of precipitation in England and Wales and a revised homogenous England and Wales precipitation series. *J. Climatol.*, **4**, 1–25.

Wilcox, J. M. Duffy, P. B. Schatten, K. H. Svalgaard, L. Scherrer, P. S. Roberts, W. O. & Olson, R. (1979). Interplanetary magnetic field polarity and the size of low pressure troughs near 180 degrees west longitude. *Science*, **204**, 60.

Williams, E. R. (1992). The Schumann resonance: a global tropical thermometer. *Science*, **256**, 1184.
    (1994). Global circuit response to seasonal variations in global surface air temperature. *Mon. Wea. Rev.*, **122**, 1917–29.

Williams, G. E. (1981). Sunspot periods in the late Precambrian glacial climate and solar–planetary relations. *Nature*, **291**, 624–8.
    (1986). The solar cycle in Precambrian time. *Sci. Am.*, August, p. 84.
    (1988). Cyclicity in the late Precambrian Elatina Formation, South Australia: solar or tidal signal? *Clim. Change*, **13**, 117–28.

Willson, R. (1997). Total solar irradiance trend during solar cycles 21 and 21. *Science*, **277**, 1963–65

Willson, R. C. & Hudson, H. S. (1991). The Sun's luminosity over a complete cycle. *Nature*, **351**, 42–4.

Willson, R. C. & Mordvinov, A. V. (2003). Secular total solar irradiance trend during solar cycles 21–23. *Geophys. Res. Lett.*, **30**, 1199–202.

Winograd, I. J. (2002). The California Current, Devils Hole, and Pleistocene climate. *Science*. **296**, 7.

Winograd, I. J., Szabo, B. J., Coplen, T. B., & Riggs, A. C. (1988). A 250,000-year climatic record from Great Basin vein calcite: implications for Milankovitch theory. *Science*, **242**, 1275–80.

Winograd, I. J., *et al.* (1992). Continuous 500,000-year climate record from vein calcite in Devil's Hole, Nevada. *Science*, **258**, 255–60.

Winograd, I. J., Landwehr, J. M., Ludwig, K. R., Coplan, T. B. & Riggs, A. C. (1997). Duration and structure of the past four interglaciations. *Quatern. Res.*, **48**, 141–54.

Wunsch, C. (2000). Moon, tides and climate. *Nature*, **405**, 743–4.

Yiou, P., Sornette, D. & Ghil, M. (2000). Data-adaptive wavelettes and multi-scale singular-spectrum analysis. *Physica D*, **142**, 254–90.

Yule, G. U. (1927). A method of investigating periodicities in disturbed series, with special reference to Wolfe's sunspot numbers. *Phil. Trans. R. Soc. Lond. A*, **226**, 267–98.

Zebiak, S. E., & Cane, M. A. (1987). A model El Niño/Southern Oscillation, *Mon. Wea. Rev.*, **115**, 2262–78.

Zhang, L. H. & Swinney, H. L. (1985). Non-propagating oscillatory modes in Couette–Taylor flow. *Phys. Rev. A*, **31**, 1006–9.

Index